UML FOR SOC DESIGN

UML for SOC Design

Edited by

GRANT MARTIN
Tensilica Inc., Santa Clara,
CA, USA

and

WOLFGANG MÜLLER
University of Paderborn,
Germany

A C.I.P. Catalogue record for this book is available from the Library of Congress.

ISBN-13 978-1-4419-3829-9 (PB)
ISBN-13 978-0-387-25745-7 (e-book)

Published by Springer,
P.O. Box 17, 3300 AA Dordrecht, The Netherlands.

www.springeronline.com

Printed on acid-free paper

Printed in the Netherlands.

Grant Martin would like to dedicate this book to his wife, Margaret Steele, and daughters Jennifer and Fiona.

Wolfgang Mueller dedicates this book to his wife, Barbara, and children Maximilian, Philipp, and Tabea.

Contents

Preface

The last several years have seen significant progress on two related fronts in hardware and software design for electronic systems. The first is the rapid growth in the design of complex System-on-Chip devices (SoC). The second is progress in adding capabilities to the Unified Modeling Language (UML) to better support the design of real-time and embedded systems, culminating in the UML 2.0 specification which is nearing final approval. It is thus an excellent time to evaluate the combination of these two topics into a unified theme: UML for SoC design.

UML 2.0 provides a collection of 13 different diagrams, which were originally targeted for application in software engineering. On the other hand, as hardware systems exceed a specific complexity, new means and methodologies are required to close the productivity gap for SoC design. UML and the closely related concept of Model-Driven Architecture (MDA) based design provide concepts, which are both of potential interest and real application for hardware design and hardware/software co-design, respectively.

At DAC 2004, we organized a UML for SoC Design workshop as a discussion forum to bring hardware, SoC, and UML experts together. For that workshop, we received great international interest and contributions from Asia, Europe, and North America. The success of the workshop has demonstrated that there is a great interest in both academia and industry to create and investigate joint efforts in SoC design and UML. This book is a collection based on the main contributors to the DAC 2004 UML for SoC Design workshop, providing the first set of papers for such a joint effort. Some additional contributions to the book were made by other experts who did not present at the workshop,

The selected chapters present approaches to executable UML, UML translations for FPGA synthesis and SystemC simulation, as well as UML-specific SoC methodologies. They give insights into the current state of the art and the most recent advances in applying UML to SoC design. They provide a representative overview of current UML activities in SoC design and give an excellent introduction to UML's application in hardware and hardware/software codesign.

We wish to acknowledge the contributions of all the contributors and participants in the 2004 DAC UML for SoC Design workshop, and the support of the DAC Executive Committee, Special Initiatives Chair, and the conference staff, without whom this book would not have been possible.

Grant Martin
Tensilica Inc., USA

Wolfgang Mueller
Paderborn University , Germany

Chapter 1

When Worlds Collide: Can UML Help SoC Design?

Grant Martin,[1] Wolfgang Mueller[2]

[1]*Tensilica, Inc.*
Santa Clara, CA, USA

[2]*Paderborn University*
Paderborn, Germany

Abstract There has been a growing realization that it is necessary to marry together notions from both the HW and SW design and verification worlds. It may be true that "HW designers are from Mars, SW designers are from Venus", and that traditionally they have not interacted where they should — but we have seen the results: late designs, poor HW-SW integration, and inadequate verification. This chapter gives an overview of how UML can help SoC design to link the worlds.

1.1 Introduction

The last several years have seen growing interest in the convergence between software and hardware design, embodied in the form of the System-on-Chip (SoC). SoC is a particularly interesting design approach for seeing this convergence, because most SoCs are intended for embedded systems, and all interesting SoCs are programmable, incorporating one or more processors. The growing trend for the future in fact is multiprocessor SoC (MPSoC) using network-on-chip (NOC) communication fabrics.

Yet the practices for design of these complex systems lag their growing complexity. There are many issues with HW design, especially in coping with advanced sub-100 nm. IC processes [43]. The complexity of these embedded systems has revitalized the area of system level design, especially using new standard languages such as SystemC [80]. But despite the fact that these systems are complex mixtures of HW and SW processing, the barriers between the traditionally divorced worlds of HW and SW are just beginning to break down. Discussions of the SW issues involved in SoC design have been occurring for

G. Martin and W. Müller (eds.), UML for SOC Design, 1–15.

a number of years [127], but the real take-off for these methods has not yet started.

1.2 New Directions for HW-SW Interaction

Designers have been looking at UML for application to embedded software and real-time design [114]. Now, SoC designers and architects have begun to look at UML for possible improvements to the specification, design, implementation and verification processes, as UML provides several means for architectural as well as behavioural design, which have been well established in HW and HW/SW Codesign for some time. The different UML diagrams or variations of them already have found application in various areas such as:

- requirements specification,
- testbenches,
- IP integration, and
- architectural and behavioural modelling.

From the general perspective of SoC design, there are three key issues and thus three opportunity areas in HW-SW integration. These are:

- Embedded system and SoC architects need to decide on their choice of programmable processors, bus architectures, memory sizes, and dedicated hardware units, and make fundamental decisions on mapping functions into hardware or software implementation. This is a traditional co-design task. With multiple processors, as in MPSoC, mapping SW tasks to particular processors, and determining how they communicate, becomes a kind of "software-software" co-design.
- HW developers need to validate the system architecture and hardware components in the context of the overall HW-SW system. In particular, Hardware-dependent Software (HdS) should be validated on transaction-accurate models of the HW system, to ensure correct HW-SW integration, long before bringing them together on the real product in the lab.
- SW developers, whether working at the HdS, middleware, or applications level, need fast-executing HW-SW platform models on which to validate the functionality of their software. In addition, at the HdS level, a transaction-accurate platform model, as described above, can also be used by SW developers for early validation.

Each of these HW-SW areas presents opportunities for ESL (Electronic System Level) tools and methods, and we have recently seen significant progress made in all of them:

- Co-design and design space exploration is supported by a number of commercial tools. Even more important, the growing community around SystemC as a standard modelling language allows the easier creation of system level models of platforms to support mapping and choices of HW or SW implementation, and a flow into implementation. These models range from functional through transaction level architectural models, and can serve the other communities of designers as well.
- For HW developers needing to validate their HW components in the context of the overall system, including the execution of HdS, the availability of recent modelling languages such as SystemC and SystemVerilog, hardware verification languages, and new verification environments which allow multiple languages to co-simulate and transaction level models from all sources to co-exist, represent substantial progress and opportunities for new methodologies. The new verification environments allow flexible creation of platform models for validating HdS in context and indeed a flow of both design and testbench models from system architects to HW designers and validation engineers is becoming a reality. The incorporation of C/C++-based Instruction Set Simulators (ISS's) into multilingual hardware verification tools is becoming standard. These new methods and flows, based on standard languages, are tending to replace the older proprietary co-simulation tools and enjoy much wider acceptance.
- For SW engineers at all levels of abstraction, we see the emergence of sufficiently accurate and fast executing functional models of processor platforms, offering them tradeoffs between transaction or functional accuracy, and speed. Such models can emerge from the system architectural and co-design arena as discussed above, or be derived from specialised processor modelling tools. Speeds of many MIPS or tens of MIPS for functionally accurate models running on standard workstations under Linux or Windows are possible, and there are an increasing number of vendors offering these models or tools to create such models.

The latest ESL developments are only a start, but a good and productive one, in offering improved productivity by allowing development, and integration of hardware and software, for platform-based embedded system design, to occur in a virtual sense at much earlier stages than has been possible with traditional methods. We are at last reaching the point where the methods, models and tools are available to address these issues very pragmatically.

1.3 UML 2.0

The Unified Modeling Language (UML) 2.0 is an Object Management Group (OMG) standard divided into four main parts: the UML metamodel infrastructure, the actual UML definition (superstructure), the XML based interchange format (XMI), and the Object Constraint Language (OCL). In this chapter, we are mainly interested in the UML definition and its application.

UML has received significant upgrades in the revision from Version 1.5 [150] to 2.0 [154] — also known as UML2. One of the most important structural updates in the UML superstructure is the definition of *components* and their ports, which represent building blocks that encapsulate structure and behaviour. Substantial upgrades in the behavioural portions introduce actions and activities derived from the concepts of the action semantics. Since the scope of the UML2 is large, and its changes from the previous UML 1.5 specification is considerable, we can only give a brief overview of UML2 and its OCL here and refer the reader interested in more details to [150, 154, 155, 156, 157, 18].

1.3.1 Structure

The fundamental structural units of UML are classes with attributes, operations, and interfaces. Class and object diagrams are the different graphical views provided for classes.

Components, represented by component diagrams, are introduced as autonomous units with interfaces exposed to other components via ports. Components have internal structure, consisting of elements that may be other components or classes, connected by *connectors*, usually through their ports. *Ports* of components are communication points, each of which may have a set of required and provided interfaces. Ports allow components to hide details of implementation (black box modelling). This concept owes much to the work of ObjecTime/Rational/IBM on capsule-port-signal notation in ROOM [192]. This method of decomposition of overall system design intent, or application structure, is more natural for most electronic system design than a purely object-oriented class-method descriptive method as offered by other parts of UML. A UML component as a structured class thus can have a realization as HW or SW, and in addition, represents the communication between components in an idealised form independent of implementation. The flexibility of components also allows use of UML for specification of behaviours using a variety of 'models of computation' (synchronous dataflow, synchronous reactive, etc.), which is an important SoC aspect especially for systems implemented on a variety of processing elements. The now classical example of a cellular handset contains control-dominated processing often implemented as SW on a RISC processor, interacting with signal processing dataflow algorithms implemented in combinations of HW and SW running on a DSP.

Run-time instances are described by composite models and their diagrams. Instances are connected by communication links with interfaces and ports, which can be adapted to represent signals over channels in a hardware context as in SystemC, or messages written into defined memory locations in a software context.

Finally, component deployments and their diagrams give a structural view of hierarchical nodes with communication paths. They basically resemble an architectural system view with execution steps on the connections and software or hardware artefacts assigned to nodes. However, they just give a rough overview of the component interaction and their expressiveness is rather limited compared to the use of the behavioural diagrams.

1.3.2 Behaviour

UML2 has use cases, actions, activities, state machines, and interactions as basic behavioural elements with various graphical representations.

Use cases represented by use case diagrams provide an informal functional view of the system and its interaction with the environment. They are typically applied to specify system requirements by means of actors, conditions, and the system as a subject of investigation. As use cases are not strongly integrated with the other behavioural parts of UML, they can be used for a variety of purposes, including application-specific uses.

The main behavioural concepts of UML combine actions, activities, and state machines, where the concept of actions was derived from the previous work of the action semantics consortium and several years of experience in generating executable code from UML.

In UML2, an action is defined as the fundamental unit of the UML behaviour. Actions can apply primitive functions, invoke behaviours, send and receive signals, and even modify structures by creating objects and links.

Actions are embedded in hierarchical and potentially asynchronous control and data flows of activities — very similar to Petri Net semantics. Activities are represented by activity diagrams. They are composed of object, control, and executable nodes connected by links defining the flow of data or control tokens. An executable node can be either an action or a structured activity. When (input/output) pins or object nodes are defined, they stand for the production and consumption of object tokens implicitly declaring a link as an object link. Those links basically compare to *places* in Petri Nets with queuing semantics and are introduced into UML2 to support the modelling of communication structures with (possibly unlimited) input and output buffers. Additional control nodes and structured activities provide advanced control flow constructs such as loop, decision, fork, and join. Activities have an exception handling but do not provide any history mechanism as supported by state machines. Activities can be grouped into partitions, which give the individual activity a context, such as hardware or a software partition or a physical location. However, the assignment has no effect on the computation.

Activities can be invoked by (hierarchical) state machines. They can be either assigned to states as entry/do/exit activities or to transitions as an effect activity. UML state machines resemble finite state machines with hierarchical (composite) and parallel (orthogonal) states. Events from activities and actions are collected in a global event pool. A dispatcher takes events from that pool and processes them in a non-defined order. For that, UML defines a run-to-completion (RTC) semantics, meaning that the processing of the next event is not started before the completion of the currently selected event. However, the RTC semantics often turns out to be a source of incompatibilities with respect to different hierarchical finite state machine implementations and Harel's StateCharts [85]. In hardware synthesis, for example, a synchronous parallel behavioural semantic is applied, which is not covered by UML. Though UML2 provides basic support for discrete-event based models of computation through activities and state machines, further investigation is necessary to apply UML as framework for models of computation. The first work in that direction is being done by SysML [207], which extends token flows by continuous token flows and probabilities.

As a complementary behavioural concept, UML introduces interactions. Interactions are partially ordered traces over event occurrences, where events are generated when a message is sent and received. Traces may have an assignment to a discrete time scale, whose relationship to any computational step of activities and state machines is not explicitly defined by the standard. Interactions can be represented as sequence, interaction overview, collaboration, and timing diagrams, where sequence diagrams owe very much to ITU's Message Sequence Charts (MSCs). As a consequence, they may be applied to various purposes such as the specification of testbenches. Sequence diagrams represent the object traces as a lifeline proceeding from the top to the bottom. Links between lifelines represent message calls. UML2 introduces new additional constructs to improve the expressive power of sequence diagrams with combined fragments covering alternatives, loops, critical sections, strict sequencing across lifelines and parallel message sequences. The decomposition of sequence diagram lifelines into hierarchical sequence diagrams allows references to embedded interactions (interaction occurrences), in order to support reuse.

1.3.3 The Object Constraint Language

The Object Constraint Language (OCL) of the UML2 is a declarative, expression based, textual language. The strength of OCL is its expressive power for the definition of expressions over collection types, mainly sets, ordered sets, bags, and sequences. For these types, OCL provides several built-in operations such as notEmpty(), includes(), includesAll(), excludes(), excludesAll(), subSequence(), exists(), forAll().

In the context of SoC design, they provide the basis for a much wider use of OCL. For example, it can be applied to the definition of temporal logic based properties for state machines which allows their use in requirement specification and formal verification such as model checking as is demonstrated in [61]. For these applications, the notion of a state has to be clearly defined in order to reason about states and state configurations. In that respect, OCL provides a much richer language than temporal logic based formulae like LTL or CTL and is more comparable to PSL/Sugar.

1.4 UML Profiles

Profiles define application-oriented variations of standard UML. A profile defines a refinement or extension of the UML metamodel typically given by class diagrams (abstract syntax) with textual outlines (class descriptions). This approach proves the applicability of UML for specific applications, since this is the process where the deep UML expertise 'collides' with domain experts. The profile document has to be readable by other domain experts, or, at least, to be translatable for their understanding.[1] To bring a profile into real application, it is required that the profile metamodel is sufficiently precise so that it can provided in a machine readable — preferably XMI based — format. Some generic modelling tools are already available and several are under development, which take such metamodels for their configurations. All in all, the definition of a profile can be a time consuming process and there is still a long way to go to arrive at precise and well-formed UML profiles for SoC design. Nevertheless, the path is already paved with several profiles relevant for SoC design; others are in preparation.

The most popular profile in the SoC community is certainly the Schedulability, Performance, and Timing Analysis (SPT) profile, which is also known as the RT profile.[2] The SPT profile provides constructs to represent more easily the kinds of timing and performance artefacts useful in embedded real-time systems, such as Rate Monotonic Analysis (RMA) and Deadline Monotonic Analysis (DMA). Though the profile is well defined and has existed for quite a while, the practices around using these constructs are not yet codified into standard approaches nor has a consensus emerged on how to make UML-based designs with constraints on real-time performance interoperable across proprietary toolsets. Such a consensus is likely to emerge in the next few years as

[1]The notion of profiles is similar to the idea of EDA information models, which came up in the early 90's and were defined by the ISO-Std10303-11 EXPRESS language [69, 113].

[2]In January 2005, the OMG initiated the development of the complementary MARTE (Modelling and Analysis of Real-Time Embedded systems) profile, a domain specific SPT extension [175].

the SPT profile and UML2 changes are supported by commercial tools and user demands for interoperability and design flow provide an imperative for standardised methods.

A related profile is the one for QoS and Fault Tolerance [158], which defines the notion of concurrently executing resource-consuming components (RCC). This profile covers real-time issues with a focus on communication policies and their latency with hard and soft deadlines. However, though the QoS profile has some overlap with the SPT profile there has been no effort to combine both on a joint basis.

An additional profile was defined for applications in software, hardware, and protocol testing. The UML testing profile [160] gives several definitions for testbenches, test architectures, stimuli, and procedures. The standard gives a mapping of the concepts to the ITU standard test language TTCN-3 (Testing and Test Control Notation), which plays an important role in telecom and automotive systems design. The mapping demonstrates its applicability for embedded systems and SoC design.

For SoC specific application, we currently see two different activities: the SoC profile and SysML. The SoC profile is an activity from the UML for SoC Forum (USoCF). The USoCF is a Japanese initiative founded by CATS, IBM/Rational, and Fujitsu to define the SoC profile with modelling guidelines. Their approach is validated by several pilot projects. The current SoC profile is based on the SPT profile and addresses the SystemC oriented SoC design process. For more details, see Section 1.4.1.

An upcoming requirement for UML comes from the need to cover a wider range of engineering applications such as automotive, aerospace, communications, and information systems. In that context, INCOSE (the International Council on Systems Engineering) devised SysML (System Modelling Language), which was submitted to the OMG as an UML profile. SysML extends UML2 by additional diagrams and concepts. One important aspect of SysML lies in its introduction of additional Models of Computation by extending the behaviour of UML activities for the modelling of continuous and probabilistic systems. Here, SysML modifies the definition of token consumption and introduces flow rates and probabilities. Additional mechanisms support the modelling of energy and material flows.

1.4.1 The UML for SoC Forum Profile

The draft proposal from the UML for SoC Forum in Japan, which was submitted to the OMG in January 2005, develops a set of extensions to allow UML to be used for SoC design [65]. A special structure diagram, the SoC structure diagram, is used in conjunction with class diagrams to create an executable system level model.

A number of stereotypes are defined to relate SoC model elements with UML metaclass elements. The essential SoC model elements include modules, connectors, channels and ports; clocks and resets; processors, protocols and data types, and controllers. As one would expect, the module is the basic class and supports hierarchy (modules contain modules and channels). Processes are member functions for behaviour. Data is communicated between modules via channels, using protocol interfaces which link to module ports. Clock channels are used for synchronisation, and reset channels control system reset behaviours.

The profile gives example of class and structure diagrams and has a defined capability for SystemC code generation that is derived naturally from the UML classes and stereotypes. In fact, the profile is arguably a translation of SystemC modelling constructs into UML terminology and diagram notation.

The SoC profile also uses OCL to define constraints on structural elements, for example, the relationships between modules, channels, ports and connectors, and interface inheritance and channel inheritance relationships. It does not yet define constraints that go beyond the structural to the behavioural domain.

1.5 Executable UML

StateCharts and their variations have been supported as a means of graphical capture by several EDA tools for more than 10 years. They have been available as graphical representations and entry mechanisms for hardware description languages such as VHDL (e.g., STATEMATE, i-Logix), together with code generation of synthesisable output. To use StateCharts as a graphical modelling front-end, states and state transitions are annotated with textual code as actions, which is very similar to the use within several CASE tools.

With the advent of UML, the notion of executable and translatable UML became a subject of wider interest and investigation [201, 131]. The notion of executable UML basically denotes the application of UML as an abstract high level programming language. In most cases, the approach to executable UML covers a well-defined UML subset in addition to a specific specification, description, or programming language (surface language). Again, in most cases, the subset covers class diagrams and state machine diagrams in combination with an application-specific programming (or hardware description) language. Each state machine typically gives the behaviour of one class with operations as its activities.

During the last few years, the notion of executable UML was also investigated within the OMG by the Action Semantics committee with the main industrial contributions from tool vendors like Project Technology,[3] Kennedy-Carter, and Kabira [1]. The Action Semantics Proposal was an additional UML package

[3] now Accelerated Technology, Mentor Graphics

that specifies computational behavior in UML 1.5 and became the behavioural basis of UML 2.0. In that context, the notion of an action language was introduced. The original motivation of the OMG for the action language was the definition of a language more abstract than a programming language with high level query and assignment constructs (cf. Mellor in [131]).

The general concepts of action semantics provided many important capabilities that enable UML to be considered as a realistic front-end specification and design vehicle for a specific application domain such as SoC design. In that context, one very promising area for future methodology work is to establish a truly multilingual world, where software specified and modelled in UML can code-generate platform-optimised software tasks, which can then be simulated within a SystemC-based transaction level platform model, with appropriate OS models. The hardware part of the design will move into SystemVerilog or VHDL based implementation and verification, re-using the functional prototypes built earlier in the process. The impetus and interest in using UML for SoC design would not have built to the levels seen recently if there were not developments in code generation that allowed both the production of executable models and the synthesis of implementation-quality code. The concepts of actions embedded in activities and state machines allow the specification of executable semantics in a syntax-neutral fashion and thus can allow a variety of specific code generators targeting various languages and platforms.

The possibilities of flexible code generation have been used in experiments with UML as a front end for hardware design, particularly targeting new system design languages such as SystemC. The convergence of interests in hardware and software development, using UML as a specification and design medium, have been accompanied by a number of experiments targeting SystemC or an HDL as an output language [182, 164, 28]. This works especially well in a codesign flow, where system function is first captured independent of realisation, and only assigned to HW or SW implementation at a later point in system analysis. A generated SystemC model can of course represent SW as well as HW. The key to this flow is to satisfy a well-defined model of interaction and communication between generated modules. Explicit communications between modules using ports, channels and interfaces as defined in SystemC is now possible in UML2 and ensures that code generation can occur cleanly. Generated processes can represent SW tasks as easily as HW modules. If one defines a specific HW-centred profile in UML, as has been done in some of the experiments, then it is possible to directly generate synthesisable Register Transfer Level (RTL) models from UML behavioural diagrams. This has been successfully demonstrated using both SystemC and HDLs.

SystemVerilog, representing recent advances in Verilog HDL, adding improved HDL constructs as well as interfaces, modules and testbench capabilities, represents another interesting target possibility for UML specifications intended to represent HW modules. Another alternative would be to generate SystemC code from UML, and refine the SystemC to a SystemVerilog form for implementation. Here, as discussed previously, the SystemC environment would be used to bring together models of SW in C/C++ or code-generated from UML, models of HW either again code-generated from UML or natively captured in SystemC, and finally, using co-simulation with HDLs, legacy HW blocks. A SystemC transaction level model of a target platform is an excellent environment within which to validate software, new hardware blocks and legacy HW blocks together.

1.6 UML in the SoC Design Process

With UML2 it is possible for designers to specify the structure and functionality of interacting components which could be realised in either software or hardware form. The interaction between components can be described as the explicit transfer of data and control tokens on communication channels — something that is more natural, and safer, for describing concurrent hardware or software behaviour than implicit communications via method calls or global variables. These channels can be ultimately realised via hardware signals or a variety of mechanisms used for software task communications — but details of implementation will be kept separate from functional modelling.

In general, the OMG notion of "Model-Driven Architecture" (MDA) fits in well with the trends over the last decade in embedded system design towards a kind of HW-SW codesign called "function-architecture codesign" [10]. In this approach, which has been widely adopted in most commercial codesign tools, the system functionality is captured in a manner separately from, and independent of, any possible architectural implementation, and can be analysed separately from any consideration of architectural effects. When designers have a notion of an explicit realisation architecture — for example, a specific target HW-SW platform — they create a mapping between function and architecture (between the "model" and the architecture), which allows analysis of the specification as a possible realisation. Estimates of performance can be compared to system constraints specified for functional processing and the communications between them.

Functional models can have behaviour specified in a number of forms — state diagrams, activity diagrams and code. UML2 provides enhanced mechanisms for behavioural specification, although it still lacks explicit support for multiple models of computation — for example, dataflow firing rules. However, the functional component modelling style can be used together with stereotyping to capture dataflow — or it can be captured as pure code.

UML provides for several mechanisms for describing component interaction, but one that has a particularly interesting potential is a combination of use case modelling and sequence diagrams for describing testcases and the generation of assertions. The ability to easily and flexibly describe expected message interactions between components, with timing and performance constraints, is one that has not yet been exploited for SoC design in details and is easily overlooked.

We will now provide additional detail on these various ideas.

1.6.1 Platform-Based Design

One particular application for UML to SoC design that has already been studied, and is highly relevant to the future application of UML in this context, is platform-based design.

The notion of a HW-SW "platform" as a design and implementation vehicle for embedded applications has been frequently discussed during the evolution of the SoC "revolution". Figure 1.1, from [37], illustrates the concept of platform-based design.

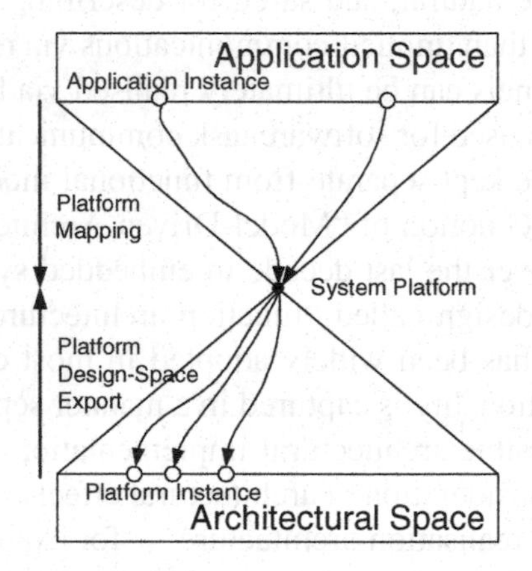

Figure 1.1. The Concept of Platform-Based Design

In Figure 1.1 we see the concept of mapping a particular embedded system application into a "system platform", which represents a series of possible platform instances from the architecture space. Platform-based design thus represents one approach to function-architecture codesign. A system platform subsumes a variety of possible platform instances whose specific configuration is explored and optimally chosen based on the notion of design space exploration.

A system platform can be described in a UML context by constructing a UML extension or profile based on adding stereotypes to the basic UML constructs. Applications, which might be described in UML collaboration diagrams, can then be mapped to this platform using these additional stereotypes. The notations of UML deployment diagrams, if extended suitably, form one basis to allow this mapping and analysis to be carried out.

Figure 1.2, again derived from the experiments in [37], illustrates the mapping between a communications protocol used in a particular application space, and an embedded system platform that is described at several layers of interface abstraction. This work, which preceded UML2 definition and standardisation, would be greatly facilitated by using the new constructs available in UML2, especially the additional semantics that facilitate the generation of executable simulation and analysis models.

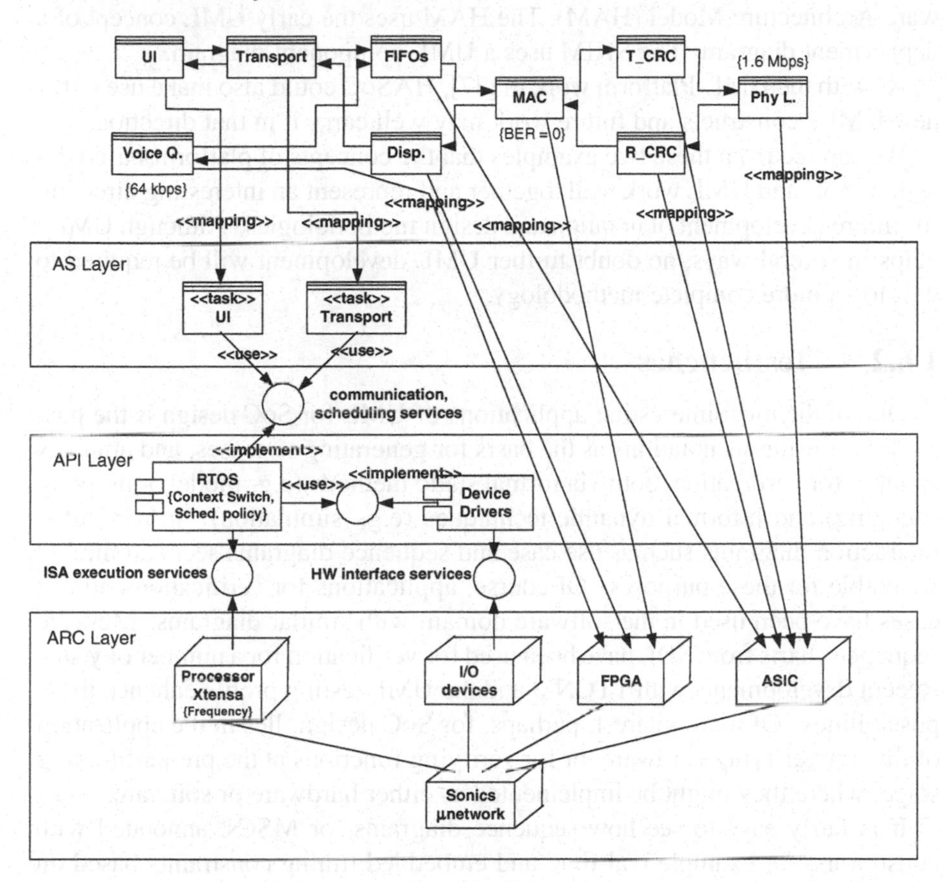

Figure 1.2. Example of Mapping a Communications Protocol onto an Embedded System Platform

Another approach to the unification of UML and SoC design is the HASoC approach of Edwards and Green [54]. In HASoC, design consists of building uncommitted applications models, and target platform models, and through a process of "committing" the application models to the platform models, (a form of mapping,) the suitability of the platform can be explored and analysed and a detailed configuration built.

Uncommitted models in HASoC are built with use cases, object and class models, state diagrams, and sequence diagrams, and executable models can be created by annotating code fragments in a manner similar to earlier action specification languages. Committing models implies HW-SW partitioning using the capsule concepts from UML-RT (that in UML2 are supported using components), and associating them with specific platform components. Platform modelling builds a Software Hardware Interface Model (SHIM) and a Hardware Architecture Model (HAM). The HAM uses the early UML concept of a deployment diagram. The SHIM uses a UML component diagram.

As with the UML-Platform work in [37], HASoC could also make use of the new UML2 constructs and future work may well carry it in that direction.

We can see from these two examples that the concepts of platform-based design of SoC and UML work well together and represent an interesting direction for future development of notation and design methodologies. Although UML2 helps in several ways, no doubt further UML development will be required to develop a more complete methodology.

1.6.2 Testbenches

One of the most interesting applications of UML for SoC design is the possibility of using its notations as the basis for generating testcases, and artefacts suitable for verification, both via formal static methods (e.g. model or property checking) and informal dynamic techniques (e.g. simulation). In particular, interaction diagrams such as use case and sequence diagrams seem admirably adaptable for these purposes. Of course, applications for verification and test cases have been used in the software domain with similar diagrams. Message sequence charts from SDL have been used for verification for a number of years. Recent developments with TTCN-3 and the UML testing profile enhance those possibilities. Of more interest, perhaps, for SoC design, lies in the application of this for verifying hardware, or for verifying functions at the pre-partitioning stage, where they might be implemented as either hardware or software.

It is fairly easy to see how sequence diagrams, or MSCs, annotated with constraints, for example real-time and embedded timing constraints based on the UML SPT profile [159], could become the basis for executable testcases. These could be expressed in a classical Hardware Description Language (HDL) such as Verilog or VHDL, or one of the new Hardware Verification Languages (HVL) such as 'e' or OpenVera, or the new hybrid HDVL SystemVerilog. Messages passed between objects as indicated in a sequence diagram can become an expected sequence of observed events, for example in a protocol timing diagram. Because sequence diagrams establish a partial ordering of messages without explicit timing (unless annotated on as constraints or SPT expressions), they can represent complex hardware or system communications protocols in a purely functional manner. Then, by adding timing or performance constraints or expressions, variably timed and fixed protocol sequences can be captured.

In addition, the new constructs allowed with sequence diagrams in UML2 such as alternatives and looping, allow more complex variable protocols to be represented. An upcoming alternative can be the use of activity diagrams for test cases, such as it is already supported by the AutomationDesk toolset from DSPACE with Python code generation [53]. The generation of executable HDL and HVL verification stimulators, monitors and checkers from sequence and activity diagrams represent one interesting set of SoC verification possibilities. Along the lines of OCL, one can also easily envisage its joint use with assertion expressions in OpenVera, SystemVerilog, e, or PSL/Sugar formats, which can be used with static formal verification tools (property or model checkers) to accomplish verification without simulation for many classes of designs. In addition, assertions can themselves be used to generate dynamic simulation verification artefacts via a number of tools. Thus, the possibilities for using UML to capture system specifications in formats useful for 'golden model verification' processes seem to be a very fruitful avenue for exploration.

Another aspect, although a minor one, of UML which can be useful in golden verification model generation is the use of Use Case diagrams to represent the span of test scenario cases which form part of the overall design specification. Use cases could be partitioned into key functional and performance cases, corner cases for special coverage, and as one means to ensure better functional coverage. They can also be used in a more rigorous development and verification process to ensure system traceability from requirements through specification through design and verification.

1.7 Conclusions

The linkage of UML and SoC design represents a set of possibilities that have to this point been experimented with, but not yet become an everyday part of designer practice. We are seeing a number of new methodologies being advocated, accompanying a tremendous burst of rapid change in design languages. Among these, the use of UML is particularly appropriate as a way of linking or bringing together the traditionally divided communities of HW and SW developers. As reusable SoC platforms become the main targets of system design, it is indeed possible that the main designers of HW in the future will be systems and software designers using UML as a specification medium, and automated flows leading them to implementations without needing deep HW design expertise. Although this is currently just a dream of a few, we have seen similar radical changes in design methods occur within our lifetimes.

Chapter 2

Why Systems-on-Chip needs *More* UML like a Hole in the Head

Stephen J. Mellor, John R. Wolfe, Campbell McCausland

Accelerated Technology, a Division of Mentor Graphics
Tucson, AZ, USA

Abstract Let's be clear from the outset: SoC most certainly can make use of UML; SoC just doesn't need more UML, or even all of it. Rather, we build executable models of system behavior and translate them into hardware and software using a small well-defined core of UML. No more UML!

2.1 Problem and Solution

2.1.1 A Caricature of the State of the Practice

In this section, we caricature today's development process so as to illuminate the problems that we address in our approach.

Partition: At the beginning of an SoC project, it is common for the hardware and software teams to build a specification, usually in natural language. This defines a proposed partitioning into hardware and software so the two teams, with different skills, can head off in parallel.

Verifying the hardware/software partition requires the ability to test the system, but it takes months of effort to produce a prototype that can be executed. Yet we need to execute the prototype before we will know whether the logic designers and the software engineers have the same understanding of the hardware/software interface. We also need to run the prototype system before we can measure its performance, but if the performance is unacceptable, we must spend weeks changing the hardware/software partition, making the entire process circular.

Interface: The only thing connecting the two separate teams, each with different skills, heading off in parallel, is a hardware/software interface specifica-

G. Martin and W. Müller (eds.), UML for SOC Design, 17–36.

tion, written in natural language. Two teams with disparate disciplines working against an ambiguous document to produce a coherent system. Sounds like a line from a cheap novel.

Invariably, the two components do not mesh properly. The reasons are myriad: the logic designers didn't really mean what they said about that register in the document; the software engineers didn't read all of the document, especially that part about waiting a microsecond between whacking those two particular bits; and of course, the most common failure mode of all, logic interface changes that came about during the construction of the behavioral models that didn't make it back into the interface specification.

Integration: So what's a few interface problems among friends? Nothing really. Just time. And money. And market share. We've been doing it this way for years. It's nothing a few days (well, weeks) in the lab won't solve. Besides, shooting these bugs is fun, and everyone is always so pleased when it finally works. It's a great bonding experience.

Eventually, the teams manage to get the prototype running, at least well enough that they can begin measuring the performance of the system. "Performance" has a number of meanings: Along with the obvious execution time and latency issues, memory usage, gate count, power consumption and its evil twin, heat dissipation, top the list of performance concerns in many of today's embedded devices.

There's nothing like a performance bottleneck to throw a bucket of cold water on the bonding rituals of the integration heroes. Unlike interface problems, you don't fix hardware/software partition problems with a few long nights in the lab. No, this is when the engineers head back to their desks to ponder why they didn't pursue that career as a long-haul truck driver. At least they'd spend more time at home.

2.1.2 Problems to Address

Given this is how we operate, what are the problems? They are, of course, deeply interrelated.

Partition: For the early part of the process, logic designers and coders are actually doing the same thing, just using different approaches. Distilled, we are all involved in:

- Gathering, analyzing, and articulating requirements for a product.
- Creating abstractions for solutions to these requirements.
- Formalizing these abstractions.
- Rendering these abstractions in a solution of some form.

- Testing the result (almost always through execution).

So, can't we all just get along? At least in the beginning.

Interface: It is typical to describe the interface between hardware and software with point-by-point connections, such as "write 0x8000 to address 0x3840 to activate the magnetron tube;" and "when the door is opened interrupt 5 will be activated." This approach requires that each connection between hardware and software be defined, one by one, even though there is commonality in the approach taken. Using a software example, each function call takes different parameters, but all function calls are implemented in the same way. At present, developers have to construct each interface by hand, both at specification and implementation time. We need to separate the kinds of interface (how we do function calls) from each interface (the parameters for each function call) and apply the kinds of interface automatically just like a compiler.

Integration: We partitioned the system specification into hardware and software because they are two very different products with different skills required to construct them. However, that partitioning introduces multiple problems as we outlined above so we need to keep the hardware and software teams working together for as long as possible early in the process by maintaining common elements between the two camps. That which they share in common is the functionality of the system—what the system does.

A significant issue in system integration, even assuming the hardware and software integrate without any problem, is that the performance may not be adequate. Moreover, even if it is adequate at the outset, version 2.0 of the product may benefit from shifting some functionality from software to hardware or vice versa. And here we are, back at the partitioning problem.

2.1.3 Towards a Solution

Happily, the problems described above do suggest some solutions.

Build a Single Application Model: The functionality of the system can be implemented in either hardware or software. It is therefore advantageous to express the solution in a manner that is independent of the implementation. The specification should be more formal than English language text, and it should raise the level of abstraction at which the specification is expressed, which, in turn, increases visibility and communication. The specification should be agreed upon by both hardware and software teams, and the desired functioning established, for each increment, as early as possible.

Build an Executable Application Model: Indeed, the specification should be executable. The UML is a vehicle for expressing executable specifications now that we have the action semantics of UML 1.5 and its latest version, the action model of UML 2.0. This action model was expressly designed to be free of implementation decisions and to allow transformation into both hardware and software. Executable application models enable earlier feedback on desired functionality.

Don't Model Implementation Structure: This follows directly from the above. If the application model must be translatable into either hardware or software, the modeling language must not contain elements designed to capture implementation, such as tasking or pipelining. At the same time, the modeling language must be rich enough to allow efficient implementations. Chief among the topics here is concurrency, both at the macro level (several threads of control as tasks or processors and blocks of logic that execute concurrently) and at the micro level (several elements in a computation executing at once).

In other words, we need to capture the natural concurrency *of the application* without specifying an implementation.

Map the Application Model to Implementation: We translate the executable UML application model into an implementation by generating text in hardware and software description languages. This is accomplished by a set of mapping rules that 'reads' selected elements of the executable UML application model and produces text. The rules, just like a software compiler generating a function call in our example above, establish the mechanisms for communicating between hardware and software according to the same pattern.

Crucially, the elements to be translated into hardware or software can be selected by marking up the application model, which allows us to change the partition between hardware and software as a part of exploring the architectural solution space.

All this provides a way to eliminate completely the hardware/software interface problems that are discovered during the initial integration, and to allow us to change the partition between the hardware and software in a matter of hours. The remainder of this chapter describes how this all works.

2.2 Executable and Translatable UML Application Models

2.2.1 Separation between Application and Architecture

A UML model can be made executable by simply adding code to it. However, this approach views a model as a blueprint to be filled out with more and more elaborate implementation detail and so ties the model to a specific implementation.

Instead, we must allow developers to model the underlying semantics of a subject matter without having to worry about *e.g.* number of the processors, data-structure organization, or the number of threads. This ability to specify functionality without specifying implementation is the difference between blueprint-type models and executable-and-translatable models.

How can we capture the functionality of the system without specifying implementation? The trick is to separate the *application* from the *architecture*, and that trick differentiates blueprint-type models from executable ones. Figure 2.1 illustrates this separation. Focus on the dotted line that separates the two parts.

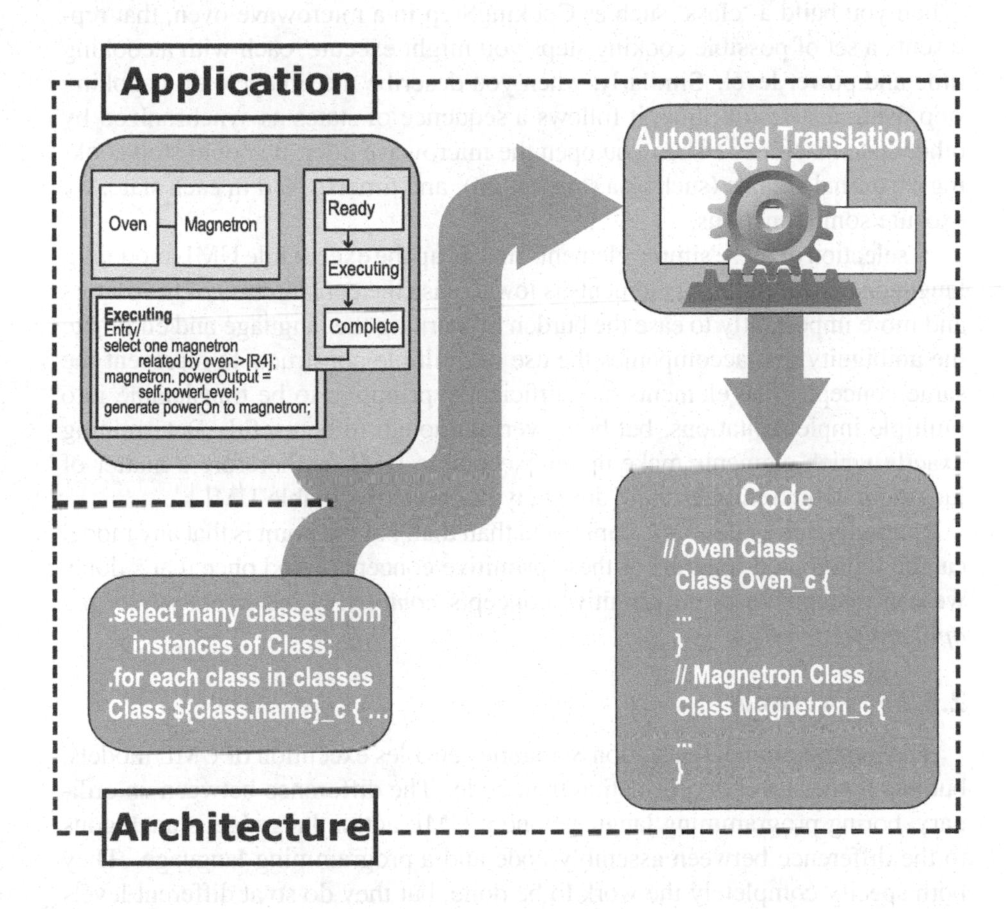

Figure 2.1. The Separation between Application and Architecture

The dotted line between "application" and "architecture" enables the capture of an executable application model by representing it in a simple form: sets of data that are to be manipulated; states the elements of the solution go through; and some functions that execute to access data, synchronize the behavior of the elements, and carry out computation. This dotted line, consisting of sets, states, and functions, captures the *modeling language*.

Only the semantics of the modeling language matter for translation purposes. If a class is represented graphically as a box, or even as text, this is of no consequence. Equally, no semantics content is added by composite structures. They can be convenient for the modeler, but to be executed they must be reduced to their constituent classes. Similar arguments can be made for the rich vocabulary of state machines that also can be reduced to simpler elements.

The UML is just an accessible graphical front-end for those simple elements. When you build a 'class' such as CookingStep in a microwave oven, that represents a set of possible cooking steps you might execute, each with a cooking time and power level. Similarly, when you describe the lifecycle of a cooking step using a state machine, it follows a sequence of states as synchronized by other state machines (when you open the microwave door, it should stop cooking), external signals (such as a stop button), and timers. And in each state, we execute some functions.

A selection of these simple elements makes up the executable UML modeling language. The number of elements is low to ease the construction of translators and more importantly to ease the burden of learning the language and eliminate the ambiguity that accompanies the use of multiple constructs to represent the same concept. The elements are sufficiently primitive to be translatable into multiple implementations, but be powerful enough to be useful. Determining exactly which elements make up an executable UML is therefore a matter of judgment, which implies there are many possible executable UMLs.

Naturally, it's a bit more complicated than that, but the point is that any model can be represented in terms of these primitive concepts. And once that's done, we can manipulate those primitive concepts *completely independently of the application details*.

2.2.2 Actions

The introduction of the Action Semantics enables execution of UML models, but at a higher level of abstraction than code. The difference between an ordinary, boring programming language and a UML action language is analogous to the difference between assembly code and a programming language. They both specify completely the work to be done, but they do so at different levels of language abstraction. Programming languages abstract away details of the hardware platform so you can write what needs to be done without having to worry about *e.g.* the number of registers on the target machine, the structure of the stack or how parameters are passed to functions. The existence of standards also made programs portable across multiple machines.

To illustrate the issues, consider the following fragment of logic:

```
Sum = 0;
Until Calls = null do
    Sum += Calls.timeOfLastCall;
    Calls = Calls.next;
endUntil
```

The elements in this fragment include assignments, `until`, `null`, an expression (with equality), an increment of a variable, and a linked list of some sort.

The semantics of `While not (expr)` and `Until (expr)` are the same. One of them is syntactic sugar—a sweeter way of expressing the same thing.[1] Of course, which one is syntactic sugar, and which the one true way of expressing the statement is often a matter of heated debate. One can always add syntactic sugar to hide a poverty of primitives *so long as the new syntax can be reduced to the primitives*. The choice of primitives is a matter of judgment and there is no bright line.

Consider, too, a triplet of classes, X, Y, and Z with associations between them. We might wish to traverse from the x instance of X to Y to Z to get the p attribute (x >Y->Z.p). The primitives involved here are a traversal and a data access. However, we could choose to implement these classes by joining them together to remove the time-consuming traversal. In this case, the specification (x->Y->Z.p) can be implemented by a simple data access xyz.p, where xyz is an instance of the combined classes X, Y, and Z. In this example, the traversal is a single logical unit expressed as a set of primitives. Defining the specification language so that these primitives are grouped together in a specification allows the translation process to be more efficient. For example, if `GetTheTimesOfAllCalls` is a query defined as a single unit in the manner of (x->Y->Z.p), so that all knowledge of the data structures is localized in one place, we can implement it however we choose.

We can go further. The logic above simply sums the times of some calls that happen to be stored in a list. We can recast it at a higher level of abstraction as: `GetTheTimesOfAllCalls`, sum them. This is the formulation used by the UML action model. The application model can be translated into a linked list as in the code above, a table, distributed across processors, or a simple array.

This last point is the reason for the word 'translatable.' An executable translatable UML has to be defined so that the primitives can be translated into any implementation. This means isolating all data access logic from the computations that act on the data. Similarly, the computations must be separated from the control structures (in the fragment above, the loop) so that the specification is independent of today's implementation—including whether to implement in hardware or software.

[1] We once saw the following code.

```
Constant Hell_Freezes_Over = False;
Until Hell_Freezes_Over do. . . . Sweet!
```

2.2.3 An Executable and Translatable UML—Static Elements

Executable and Translatable UML (xtUML, or just Executable UML) [131] defines a carefully selected streamlined subset of UML to support the needs of execution- and translation-based development, which is enforced not by convention but by execution: Either an application model compiles and executes correctly, or it doesn't.

The notational subset has an underlying execution. All diagrams (*e.g.* class diagrams, state machines, activity specifications) are "projections" or "views" of this underlying model. Other UML models that do not support execution, such as use case diagrams, may be used freely to help build the xtUML models, but as they do not have an executable semantics, they are not a part of the xtUML language. The xtUML model is the formal specification of the solution to be built on the chip.

The essential components of xtUML are illustrated in Figure 2.2, which shows a set of classes and objects that use state[2] machines to communicate. Each state machine has a set of actions triggered by the state changes in the state machine. The actions then cause synchronization, data access, and functional computations to be executed.

Each class may be endowed with a state model that describes the behavior of each instance of the class. The state model uses only states (including initial pseudostates and final states), transitions, events, and actions. Each class may also be endowed with a state model that describes the behavior of the collection of instances. Each class in a subtyping hierarchy may have a state model defined for it as a graphical convenience.

A complete set of actions, as provided by UML 1.5 and later, makes UML a computationally-complete specification language with a defined "abstract syntax" for creating objects, sending signals to them, accessing data about instances, and executing general computations. An action language "concrete syntax" [3] provides a notation for expressing these computations.

2.2.4 xtUML Dynamics

Figure 2.2 showed the static structure of xtUML, but a language is meaningful only with a definition of the dynamics. To execute and translate, the language has to have well-defined execution semantics that define how it exe-

[2]UML 2 uses "state machine" to mean both the diagram (previously known as a state chart diagram) and the executing instance that has state. To reduce ambiguity, we have chosen to use "state model" for the diagram describing the behaviors, and "state machine" for the executing instance. Where either meaning could apply, we use "state machine" to be consistent with the UML.

[3]BridgePoint ® provides OAL (Object Action Language) that is compliant with the abstract syntax standard, but there is presently no action language (notation) standard.

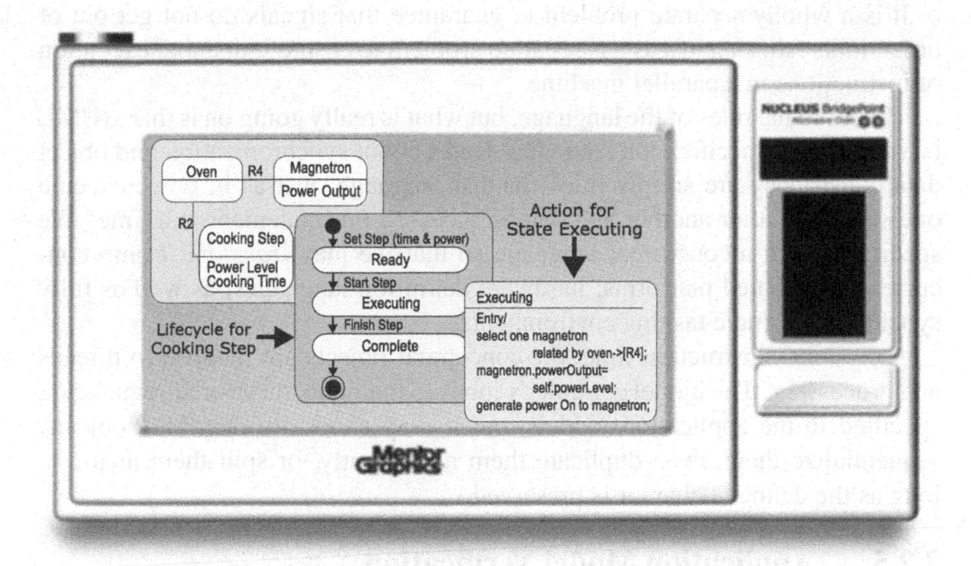

Figure 2.2. The Structure of an xtUML Model

cutes at run time. xtUML has specific unambiguous rules regarding dynamic behavior, stated in terms of a set of communicating state machines, the only active elements in an xtUML program.

Each object and class can have a state model that captures its behavior over time. Every state machine is in exactly one state at a time, and all state machines execute concurrently with respect to one another. Each state machine synchronizes its behavior with another by sending a signal that is interpreted by the receiver's state machine as an event. On receipt of a signal, a state machine fires a transition and executes an activity, a set of actions that must run to completion before that state machine processes the next event.

Each activity comprises a set of actions that execute concurrently unless otherwise constrained by data or control flow, and these actions may access data of other objects. It is the proper task of the modeler to specify the correct sequencing and to ensure object data consistency.

The essential elements, then, are a set of classes and objects with concurrently executing state machines. State machines communicate only by signals. On receipt of a signal, a state machine executes a set of actions that runs to completion before the next signal is processed. (Other state machines can continue merrily along their way. Only *this* state machine's activities must run-to-completion.)

Signal order is preserved between sender and receiver object pairs, so that the actions in the destination state of the receiver execute *after* the action that sent the signal. This captures desired cause and effect.

It is a wholly separate problem to guarantee that signals do not get out of order, links fail, *etc.*, just as it is separate problem to ensure sequential execution of instructions in a parallel machine.

Those are the rules of the language, but what is really going on is that xtUML is a concurrent specification language. Rules about synchronization and object data consistency are simply rules for that language, just as in C we execute one statement after another and data is accessed one statement at a time. We specify in such a concurrent language so that we may *translate it* onto concurrent, distributed platforms; hardware definition languages; as well as fully synchronous, single tasking environments.

At system construction time, the conceptual objects are mapped to threads and processors. The model compiler's job is to maintain the desired sequencing specified in the application models, but it may choose to distribute objects, sequentialize them, even duplicate them redundantly, or split them apart, so long as the defined behavior is preserved.

2.2.5 Application Model Verification

An application model can be executed independent of implementation. No design details or code need be added, so formal test cases can be executed against the application model to verify that requirements have been properly met. Critically, xtUML is structured to allow developers to model the underlying semantics of a problem without having to worry about whether it is to be implemented in hardware or software.

Verification of application models is exactly that: It verifies the behavior of the application models and nothing else. It does not check that system will be fast enough or small enough; it checks only that the application does what you (your client and your experts) want. Verification is carried out by model execution.

When we construct an application model, such as that shown in Figure 2.2, we declare the *types* of entities, but models execute on instances. To test the behavior of the example in Figure 2.2, therefore, we need first to create instances. We can do this with action language or with an instance editor.

We then define operations to create the object instances on which the test will execute. Then, we send a signal to one of the created objects to cause it to begin going though its lifecycle. Of course, actions in that object may be signal-sends that cause behavior in other objects. Eventually the cascade of signals will cease and the system will once again be in a steady state, ready for another test.

Each test can be run interpretively. Should the test fail, the application model can be changed immediately and the test rerun. Care should be exercised to ensure the correct initial conditions still apply. It is useful to be able to 'run' the whole test at once; to 'jog' through each state change, and to 'step' through each action.

When each test run is complete, we need to establish whether it passed. This can be achieved either by interactively inspecting attribute values and the state of the application models or by using action language that simply checks expected values against actuals. A report can be produced in regression-test mode that states whether each test passed or failed.

Selecting test cases is a separate topic. There are many theories on how best to test a system ranging from the "just bang on it till it breaks" end of the spectrum through to attempts at complete coverage.

2.3 Manipulating the Application Models

2.3.1 Capturing Application Models

When we build an application model such as that in Figure 2.2, its semantic content must be captured somehow. This is accomplished by building a model of the modeling language itself.

The classes Oven and CookingStep, for example, are both instances of the class Class in the model of the modeling language[4]. Similarly, the states Ready and Executing for the class CookingStep are captured as instances of the class State, and attributes powerOutput and pulseTimer of the class Magnetron are captured as instances of the class Attribute. This is also true for the actions. The action language statement `generate powerOn to magnetron;` is an instance of the metamodel class GenerateSignalAction.

Naturally, a tool will also capture information about the graphical layout of the boxes on the diagrams entered by the developer, but this is not necessary for the translation process. Only the semantics of the application model is necessary for that, and that's exactly what is captured in a metamodel.

2.3.2 Rules and Rule Languages

We have captured the semantics of an application model completely in a neutral formalism allowing us to write rules. One rule might take a 'class' represented as a set CookingStep(cookingTime, powerLevel), and produce a class declaration. Crucially, the rule could just as easily produce a struct for a C program, or a VHDL entity. Similarly, we may define rules that turn states into arrays, lists, switch statements, or follow the State pattern from the Design Patterns community. (This is why we put 'class' in quotation marks. A 'class' in an executable model represents the set of data that can be transformed

[4] The formal name for a model whose instances are types in another model is a *metamodel*.

into anything that captures that data; in a blueprint-type model, a class is an instruction to construct a class in the software.)

These rules let us separate the application from the architecture. The xtUML model captures the problem domain graphically, and represents it in the metamodel. The rules read the application as stored in the metamodel, and turn that into code.

There are many possible rule languages. All that's required is the ability to traverse a model and emit text. As an example, the rule below generates code for private data members of a class by selecting all related attributes and iterating over them. All lines beginning with a period ('.') are commands to the rule language processor, which traverses the metamodel whose instances represent the executable model and performs text substitutions.

```
.Function PrivateDataMember( class Class )
.select many PDMs from instances of Attribute related to Class
.for each PDM in PDMs
${PDM.Type} ${PDM.Name};
.endfor
```

`${PDM.Type}` recovers the type of the attribute, and substitutes it on the output stream. Similarly, the fragment `${PDM.Name}` substitutes the name of the attribute. The space that separates them and the lone ';' is just text, copied without change onto the output stream.

Table 2.1. C++ Code Generation

Rule	**Generated Code**

.select many *stateS* related to instances of
    class->[R13]StateChart->[R14]State
        where (isFinal == False);
public:
    enum states_e
        { NO_STATE = 0,
.for each *state* in *stateS*
    .if (not last *stateS*)
        ${*state*.Name},
    .else
        NUM_STATES = ${*state*.Name}
    .endif;
.endfor;
        };
	
	public:
	    enum states_e
	    { NO_STATE = 0,
	      Ready,
	      Executing,
	      NUM_STATES =
	          Complete
	    };

In the more complete example in Table 2.1, the rule uses italics for references to instances of the metamodel; underlining to refer to names of classes and attributes in the metamodel; and noticeably different capitalization to distinguish between collections of instances vs. individual ones.

You may wonder what the produced code is for. It is an enumeration of states with a variable `NUM_STATES` automatically set to be the count for the number of elements in the enumeration. (There is a similar rule that produces an enumeration of signals.) The enumerations are used to declare a two-dimensional array containing the pointers to the activity to be executed. You may not like this code, or you may have a better way to do it. Cool: all you have to do is modify the rule and regenerate. Every single place where this code appears will then be changed. Propagating changes this way enables rapid performance optimization.

While the generated code is less than half the size of the rule, the rule can generate any number of these enumerations, all in the same way, all equally right—or wrong.

We have also used the rule language to generate VHDL in Table 2.2.

Table 2.2. VHDL Code Generation

Rule	**Generated Code**

`.select many stateS related to instances of`
`    class->[R13]StateChart->[R14]State`
`        where (isFinal == False);`
`TYPE t_${class.Name}State IS {`
`.for each state in stateS`
`    .if (not last stateS)`
`        ${state.Name},`
`    .else`
`        ${state.Name}`
`    .end if`
`.end for`
`};`
	`TYPE t_CookingStepState`
	`IS {`
	`    Ready,`
	`    Executing,`
	`    Complete`
	`};`

The rule language can be used in conjunction with the generator to generate code in any language: C, C++, Java, Ada, and, if you know the syntax, Klingon.

2.3.3 Model Compilers and System Construction

So, we can build a platform-independent model of our application, we can execute it to verify that it functions properly, and as we've just seen we can translate it into text of any form. The instrument for organizing the collection of translation rules is a model compiler, and as a result, the overall architecture of the generated system is then encapsulated within the model compiler.

Each model compiler is coupled to the target, but the model compiler is independent of the application models that it translates. This is important because maintaining this separation of application from design allows us to reuse both application model and model compiler as needed. We can translate the same application model for many different targets by using different model compilers, but the models of the application do not change. Similarly we can use the same model compiler to translate any number of application models for a given target without changing the model compiler.

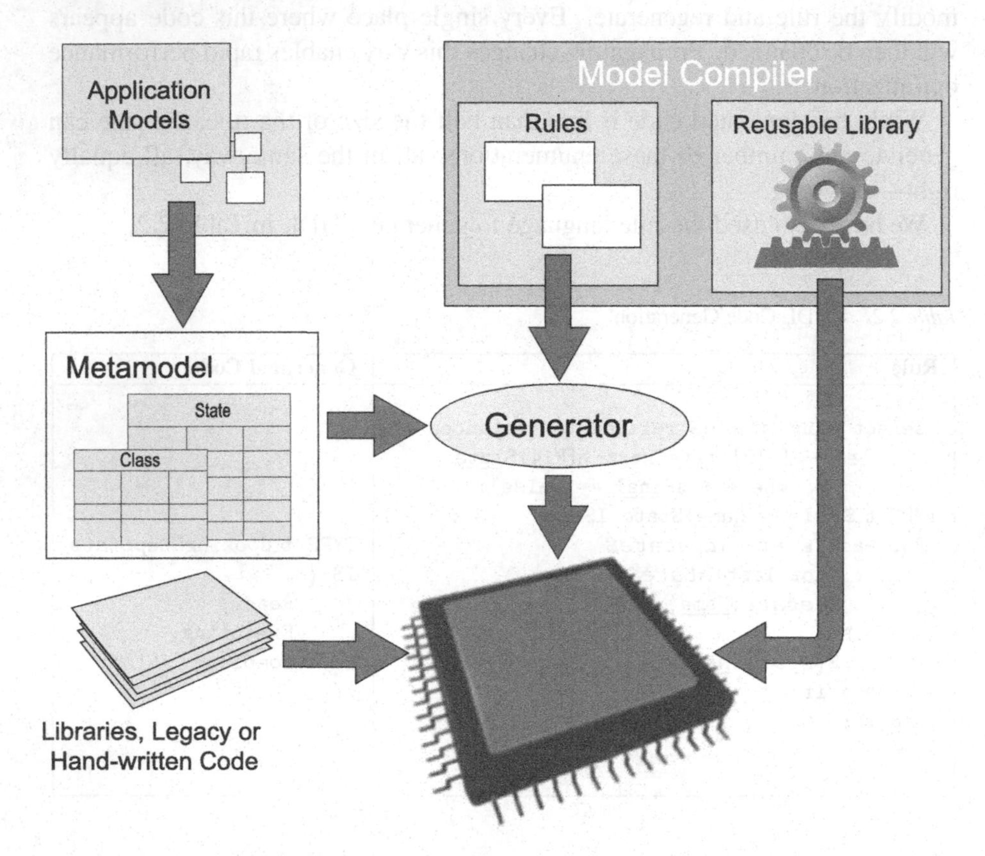

Figure 2.3. Model Compiler and System Construction

Each model compiler encapsulates the architecture of the generated system. If we're generating only software, then a software architecture—the design and implementation approaches used to render each element of the model in some implementation language like C++ or C—has been captured.

For example, a model compiler for an object-oriented architecture will likely translate each UML class to a C++ class or a C struct with each associated UML attribute being translated to a data member of the class or a member of the struct.

The state machines in the application model would be translated into two dimensional arrays where the row index represents the current state of an object, and the received event provides the column index. Each cell contains the value of the next state for the transition identified by the current state (row) and the received event (column). This next-state value is then used as an index into an array of function pointers, each corresponding to a state. This particular approach leads to a constant-time dispatching mechanism for the actions of each state machine.

Of course we can use alternative implementations depending on our needs. For example in some cases, we might choose to use a switch statement or nested if-else statements to implement the state machine, each of which would have slightly different speed and space characteristics (if-else is linear in the product of states and events).

For hardware implementations we might choose to translate each UML class into a collection of registers, one for each attribute in the class. Each state machine of the application model could be mapped to a VHDL case statement. There are, of course, many other possible implementations. UML classes can be mapped into blocks of RAM, and state machines can be translated into a data-driven and gate-conservative dispatcher.

But what about the interfaces between the hardware and software components of the system? These interfaces are just a part of the architecture encapsulated within the model compiler. Let's look at a simple example.

Suppose we have two UML classes, CookingStep and Oven where CookingStep is translated to software and Oven is translated to hardware. In this case the hardware architecture for Oven is a collection of memory-mapped registers. The generated interface for an action in CookingStep that accesses an attribute of Oven is then a simple memory access of the address allocated for the register in question.

Consider a slightly different hardware architecture in which the UML class Oven is mapped to only two addresses, one representing a control register and one for a data register. Accesses to attributes of the class would then be accomplished by writing the attribute identifier and the type of access (read or write) to the control register and then manipulating the data register accordingly (reading it after writing to the control register or writing it before writing to the control register).

The point here is that since the model compiler is generating the implementation for both the hardware and the software components of the system, it has sufficient information to generate the interface between the two, and it will do so without succumbing to the ambiguities of a natural language specification. It will do it correctly, every time.

It's possible to build and deploy model compilers that provide completely open translation systems. Exposing the translation rules in this way allows you to make changes to the rules to meet your particular needs so that the model compiler generates code exactly the way you want. This puts the translation of the models completely under your control.[5]

OK, we have collections of rules that translate UML models into hardware and software implementations. In many cases we have multiple rules for translating each UML construct, one rule for each variation in our architecture. Now we need a way of selecting which rule to apply to which part of the model.

2.4 Marks

We have described a model compiler that has two parts, but we have not yet described how we tell the model compiler whether to generate hardware or software for a given model element. To do so, we need additional inputs to decide which mapping to perform. These additional inputs are provided as *marks*, which are lightweight, non-intrusive extensions to models that capture information required for model transformation without polluting those models.

Each kind of mark has a name and a type. In addition, a kind of mark can have a default value. In programmer-esque language, we might write

Mark HardSoft [isHardware, isSoftware] = isSoftware

which declares a kind of mark named HardSoft that can have one of two values, where the default is isSoftware.

Most marks apply to instances of metaclasses, so that, for example, if we have a metamodel with class Class, and two instances of that class, Oven and CookingStep, the mark HardSoft can have a separate value for each of those instances, isHardware for the Oven, say, and isSoftware for the CookingStep.

Were the kind of mark to apply instead to generated signals, then the marks would be associated with instances of the class GenerateSignalAction in the metamodel. A given application model element can have several marks associated with it. Each of these marks is an extended attribute of the appropriate metamodel class.

We do not intend to leave the impression that the metamodel should be extended directly. Marks are not part of either the application model or the metamodel, though they can refer to them both. Rather, we view the extended attributes of the metaclasses as being housed on a plastic sheet that can be peeled off at will for a different model compiler. This separation supports both model portability and longevity.

The separation also provides the ability to evaluate a number of different architectural possibilities without requiring modification of the application model. Just change the values of the marks. Not to put too fine a point on it, this solves

[5] We know of at least one vendor providing commercial model compilers in source-code form. More are sure to follow suit in the coming years.

the problem of changing the hardware-software partition after we have verified the behavior of the application model by executing test cases against it.

The plastic sheet analogy suggests that some marks might be related and could all be placed on the same sheet. A single sheet could contain multiple related marks, such as those indicating which types of hardware implementations should be applied to which elements of the application model. Removal of the plastic sheet, then, implies the removal of the entire hardware architecture represented by that sheet from the system.

Marks may also be quantities used to optimize the target implementation. Consider a model that must be transformed into an implementation that occupies as small an amount of memory as possible. We can save memory if we observe that a particular class has only one instance (*e.g.*, Oven). Such a class can be marked as extentLess, and no container need be generated for instances of that class, making references to the instances more efficient. Similarly, we can make trade-offs within the hardware architecture. Suppose the original target had plenty of address space available and consequently mapped each class attribute implemented in hardware to a specific address, making the software access to the attributes simple and efficient. In a subsequent release we move to a lower-cost processor with a constrained address space. Through marks we then instruct the model compiler to use a single pair of addresses for each class to provide access to all the attributes in the class. Since the model compiler knows how to generate the software required for this more interesting approach for accessing hardware-resident attributes, the application models do not change even though the nature of the hardware/software interface has been drastically altered.

There have to be ways for the modeler to assign values to marks. Some implementations provide for graphical drag-and-drop allocation of model elements into folders that correspond with marks; others define an editor for the defined mark sets that can display all marks defined by the model for a selected model element, with pull-down menus for each of the marks. Another option is to define text files, and then use the large set of available editing, and scripting tools.

2.5 Work in Context

The work described here, UML models, metamodels, and transformations to text all fit into a larger context within the Object Management Group. The OMG, the organization that standardized the UML, has a standard approach for storing metamodels, not just the UML metamodel, and it is working now [152] to define a standard approach for transforming populated models to text.

These standards, and others, fit into a larger architecture called Model-Driven Architecture. You may have heard of MDA in an IT context, but the principles behind it apply to system development in general, and they're not specific to a certain kind of system or even to software [132].

MDA is three things:[6]

- An OMG initiative to develop standards based on the idea that modeling is a better foundation for developing and maintaining systems
- A brand for standards and products that adhere to those standards
- A set of technologies and techniques associated with those standards

At present, MDA is still in development, and some of the technologies need further definition while others need to be standardized.

One key standard that is missing at the time of writing is a standard definition of an executable, translatable UML. While there are mechanisms that allow for the interchange of models between tools, there must be agreement on the UML elements they can each understand. That is, the tools must share a common subset of UML for the tools to communicate effectively. It is possible for one tool to be "more powerful" than another, but effectively that restricts the power of a two-tool tool chain to the weaker of the two. When multiple tools claim to be in the same tool chain, it is only as strong as its weakest link.

Model-driven *architecture*, of course, is the name of the game. Not only must there be standards for the UML and the rest, but it's also important that tools built to these standards also fit together within that architecture and so create a complete model-driven development environment. This set of tools, loosely sequenced, constitutes a tool chain.

There are many possible tools that need to be integrated to make a complete development environment. With the right standards, one can envision tools that do the following:

- Transform one representation of an underlying model to another representation friendlier to a reader
- Generate test vectors against the application model, and then run them
- Check for state-space completeness, decidability, reachability, and the like
- Manage deployment into processors, hardware, software, and firmware
- Mark models

[6]Dr. Richard Mark Soley, the Chairman of the OMG, defined MDA thusly. We also used his definition in [132], for which Dr. Soley was a reviewer.

- Partition or combine behavior application models for visualization or deployment
- Analyze performance against a given deployment
- Examine the generated code in execution (in other words, model debuggers)

We can imagine a developer receiving a application model from a vendor or colleague; turning that application model into a comfortable notation or format; making a change with a model builder; verifying that the behavior is correct by analysis and by running test cases; marking the models and deploying them; analyzing performance; debugging the resulting system, and so forth.

When developers have the ability to provide specification tidbits at varying levels of abstraction and then link them all together, MDA will face additional tool challenges regarding smooth integration between different specification levels, such as consistency checking, constraint propagation, and incremental mapping-function execution.

As the SoC community continues to push toward a higher level of abstraction for the specification of systems it will be important to establish and maintain relationships with organizations focused on model-driven development, even if the traditional focus of such groups has been software development.

2.6 How Does All This Stack Up?

Partition? Because the xtUML models accurately and precisely represent the application, and the implementation is generated, with absolute fidelity from these models, the partition can easily be changed, and a new implementation can then be generated. This replaces weeks of tedious manual changes to an implementation with a few hours of automatic generation.

With the ability to change the partition and regenerate the implementation, the developers can explore much more of the design space, measuring the performance of various allocation arrangements that would otherwise be prohibitively expensive to produce.

Interface? The interface between the hardware and software is defined by the model compiler. Because the implementation is generated, there can be no interface mismatches. Because we no longer have two separate teams of people working from a natural-language interface specification, the generated implementation is guaranteed to have exactly zero interface problems. (This does have the unfortunate side effect of reducing the number of opportunities for logic designers and software engineers to spend long nights together in the integration lab fixing interface problems.)

Marks are a non-intrusive way to specify allocation of system functionality without affecting system behavior. The existence of automated tools to cause the generation of the system with interfaces known to be correct completely removes any interface problems that can lead to a failure to integrate the system.

Integration? Integration is now the integration of two independently tested pieces: the application model and the model compiler. Each can be understood separately, tested separately, reviewed separately, and built separately. The integration is realized by the generalized part of the model compiler (the generator) that embeds the application into the target platform.

The real integration issue now whether that model compiler meets the performance needs of the system. Should performance be less than adequate, we can change the allocation using the existing model compiler, or change the rules to create a new target architecture that does meet the performance requirements

2.7 A Hole in the Head?

SoC needs UML, but not a lot of it, and even less does SoC need more UML, especially more UML used specifically to capture hardware implementation. That way lies the primrose path to the ever-burning fires.

All SoC needs is a small, but powerful, subset of UML enabling abstract specification of behavior. Automated mappings enable interface definition in one place, so that consistency is guaranteed. Marks enable late decision making on the partition.

That's all we need; we need more UML like a hole in the head.

Chapter 3

UML as a Framework for Combining Different Models of Computation

Peter Green

School of Electrical and Electronic Engineering
University of Manchester
Manchester, United Kingdom

Abstract This chapter discusses how the well-known synchronous dataflow formalism can be represented in UML 2.0, and how it can be integrated with UML state machines to provide an object oriented specification language encompassing both state-based and dataflow behavior.

3.1 Introduction

Rising silicon capacity underpins the development of embedded systems-on-chip (SoCs) with increasingly complex behavior [34], and consequently the specification of system behavior becomes progressively more challenging. Behavior is typically classified as control- or data-oriented, and the different types of behavior are normally specified by different models of computation [195]. However, since it is the complete behavior of a system that meets requirements, then in order to assess a specification, for example by execution, it is necessary to develop frameworks in which different models of computation can be combined, or to invent new formalisms [195, 68].

The purpose of a model of computation is to support the specification of system behavior, and to facilitate behavioral analysis. Such formalisms are not intended to represent the overall organization, or structure, of a system. However, increasing silicon capacity leads to increasing structural complexity, and so notations to support the description of structure are required. The need for a comprehensive and coherent set of modeling notations to support the specification of both behavior and structure has led the software community to adopt UML as the *de facto* standard modeling language for software-intensive sys-

G. Martin and W. Müller (eds.), UML for SOC Design, 37–62.

tems, since it provides a rich set of constructs to support both types of modeling. The success of UML within the software community has prompted researchers in the embedded SoC field to investigate the applicability of UML in this area [126, 78]. However, a number of limitations with respect to its applicability to embedded SoCs have been identified [126, 78], prompting researchers to propose the development of a specialization (or profile) of UML suitable for complete SoCs, including software and hardware, application and platform.

One difficulty with using UML for embedded SoCs is that although it provides extensive facilities for the modeling of system structure, and a rich set of constructs for modeling control-oriented behavior, until recently it has not provided support for the high-level modeling of dataflow behavior, which is common in DSP and video applications. However, the enhanced activity model in UML 2.0 enables such behavior to be represented. Hence the purpose of this chapter is to demonstrate how UML can be used as a framework for integrating the specification of control and dataflow behavior, and how implementations, which may or may not use object oriented (OO) techniques, can be derived. The synchronous dataflow formalism is used to represent the behavior of classes and objects that exhibit dataflow behavior. It was selected because it is well-known within the embedded systems community, and significant experience has been gained in its application [115]. It also has a well-developed mathematical basis.

Hence the overall intention is to develop an OO framework for the specification of embedded SoCs that integrates both state-based and dataflow behavioral models. The result is a system that is specified as a network of objects, some of which are reactive, some of which continuously process streams of data, and others which share both characteristics. The work is related to other research that has sought to integrate different models of computation; in particular, it draws on [68] and [10].

3.2 Modeling Framework

This work takes place in the context of the HASoC development method [78], although the formalism that has been developed may also be used in other design flows. HASoC provides a framework for developing both the software and hardware of an embedded SoC, and both application and platform are modeled in OO terms with UML. Briefly, the HASoC method is an iterative, incremental approach to development, merging concepts from the MOOSE method [136], the Rational Development Lifecycle [24] and platform-based design [34]. Within HASoC, the approach to application development is use case driven, involving the iterative development of class and object models that realize increasing numbers of system use cases as development proceeds. Each version of the object model is itself subjected to an iterative process of *commitment*, where the object model is partitioned between software and hardware, and mapped to the current version of the underlying platform for evaluation in terms of the non-functional characteristics of interest. This may result in the modification of the object model, the partition or the platform.

Part of the development of class/object models involves the specification of internal behavior. This can be described in terms of some programming language or hardware description language, or in terms of a particular model of computation (MoC). A MoC is typically restricted in its expressiveness compared with an *ad hoc* description, but provides benefits in terms of support for formal or semi-formal reasoning, partitioning and the automatic synthesis of an implementation. For these reasons, models of computation are the principle method of representing the behavior of classes/objects within HASoC, although developers may also supply arbitrary code sequences. UML state machines are used for modeling control-oriented behavior, and synchronous dataflow (SDF) graphs are used to represent dataflow components. SDF graphs are not part of UML, but may be described using activities in UML 2.0, as discussed in Section 3.3.3. Hence object models in HASoC consist of sets of collaborating entities, some of which contain state machines, some of which contain SDF graphs, and some of which contain both types of behavioral description. Modeling systems in this way raises issues of communication and concurrency, and these are discussed in Section 3.4. For objects that contain both state machines and SDF graphs, the way in which the two types of behavior can interact must be considered, and this is discussed in Section 3.7.

3.3 Modeling Synchronous Dataflow Graphs in UML 2.0

This section reviews salient aspects of the notation and semantics of SDF graphs and activities in UML 2.0. There then follows a discussion of how SDF graphs may be represented by UML activities.

3.3.1 The Synchronous Dataflow Model

The SDF formalism was introduced by Lee and Messerschmitt [116] as an approach to modeling signal processing applications and supporting automatic code generation for DSPs. It has undergone continuous refinement since its introduction, and has been widely used [115]. In SDF, a computation is represented as a graph, the nodes, or *actors*, representing sub-computations and the arcs representing communication paths between the actors. A path between two actors explicitly represents data dependencies between the two. Consequently, actor computations must be free from side-effects.

The SDF model is data-driven, meaning that a node may perform its computation (fire) whenever data values (tokens) are available on all of its input arcs. When a node fires, it consumes tokens from each of its input arcs and deposits new tokens on its output arcs, which represent the result of its computation.

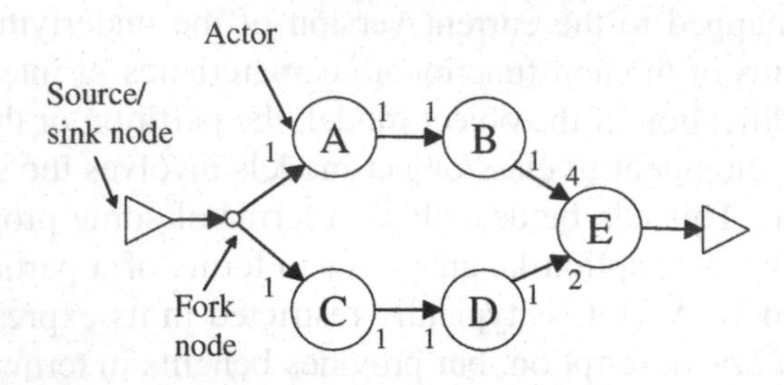

Figure 3.1. Simple SDG Graph (adapted from [16])

Tokens are stored in FIFO order in buffers associated with the arcs. The numbers of tokens consumed from each input arc and produced on each output arc are fixed. Since data availability alone controls the firing of actors, then it is possible that several actors may fire simultaneously, and consequently the SDF model can exploit the full concurrency available within an algorithm.

Figure 3.1 shows a simple SDF graph. The circular nodes represent actors, and the triangular nodes represent sources and sinks of tokens i.e. the interface between the SDF graph and its environment. The small circular, unlabelled nodes are fork nodes. These are a notational convenience, and replicate inputs on each of the output arcs.

Actors may have a range of granularities, ranging from the fine where, for example, the actor implements a simple arithmetic operation, to the coarse, where an actor can represent a significant DSP operation. In the latter case, an actor may itself be represented by a lower-level SDF graph, and so models may be hierarchical. Fine grained actors that are not refined are said to be *atomic*.

The basic SDF model involves the streaming of a semi-infinite sequences of samples through the SDF graph. Actors fire in an order that respects the data dependencies implied by the node edges. Since semi-infinite sample sequences are being processed, the schedule of actor firings is periodic. Lee and Messerschmitt [116] demonstrate that for SDF graphs with consistent sample rates, a schedule can be computed at design-time that will execute in bounded memory. Sequential and concurrent implementations can be derived in software or hardware [16, 229].

3.3.2 Activities in UML 2.0

The UML 2.0 activity model provides facilities for describing behavior in terms of control and data flows [154] and may be applied in areas as diverse as algorithms and business processes. Activity diagrams can be used to represent an operation of a classifier, the overall behavior of a classifier, the behavior of use cases without reference to classifiers, or just abstract behavior, independent of classes/objects.

An activity is a directed graph that describes how actions are sequenced within the behavior. An action is an atomic unit of behavior, and the UML 2.0 standard defines several different types of action. In the context of representing SDF graphs, two types are relevant: computation actions and call actions, and these will be considered in the next section.

Activities may have other types of nodes, besides actions. Control nodes are used when there are multiple possible paths through the activity, and can represent decision and merging points, the forking and joining of concurrent flows, and the initial and final steps within the activity.

Edges within an activity are used to transport objects or values (object flows) or control (control flows). Hence, the basic model of an activity concerns values, or objects, flowing along the paths through the activity, being modified or transformed as they pass through actions, and being routed by control nodes.

Another type of node, the object node, may be used to temporarily store object tokens/values, and such nodes have an upper bound that defines their buffering capacity. There are two special types of object node: pins and activity parameter nodes [21]. Pins are used to connect input and output object flows to actions, and parameter nodes are used to represent the parameters that are passed into and out of an activity.

The execution semantics of activities is defined in terms of tokens flowing through the activity. In abstract terms, tokens transport objects or values over the edges between nodes. Token movement can also represent the flow of control through a system, in a similar fashion to Petri Nets. Actions may only execute when there are tokens present at *all* of their inputs, both object and control flows. Lower and upper bound multiplicity values may be associated with input and output pins, specifying the minimum and maximum numbers of tokens required or provided when the action fires.

Object flows between source and destination nodes have an associated weight that determines *when* tokens are transferred from the output pin of the source to the input pin of the destination. Consider an action that produces m tokens per invocation and which is connected to a destination node via an arc of weight n. Further assume that $m < n$. Every time the source node fires it generates m tokens, to add to the running total of tokens t generated by previous invocations. Tokens are only transported over the arc when $t \geq n$, and then n tokens will be transferred.

An example activity diagram is shown in Figure 3.2.

3.3.3 Representing SDF Graphs with Activity Diagrams

UML does not support any of the dataflow formalisms typically used in the development of embedded SoCs. However, this section presents a discussion

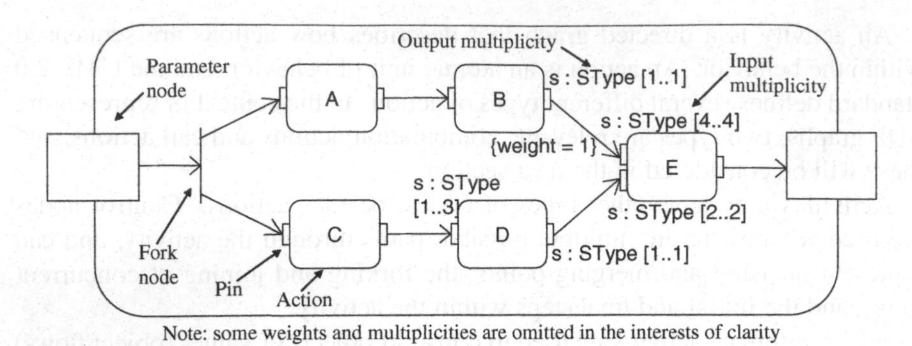

Figure 3.2. Activity Diagram Corresponding to the SDF Graph of Figure 3.1

of how activities in UML 2.0 may be used to represent the semantics of SDF graphs.

A comparison of the SDF and UML activity models reveals broad similarities. Both emphasize the flow of tokens bearing information through a system, and both consider token presence as the trigger to node execution. With its rich set of constructs, the UML activity model is applicable to a far wider range of systems than the SDF model. However, the SDF graph model's simplicity offers a number of significant advantages. Concurrency may be exploited more readily since SDF graphs can be statically (i.e. at design-time) and automatically scheduled. This enables concurrent implementations in software or hardware to be automatically generated, with sub-graphs being mapped to different execution units. In addition, consistent SDF graphs can be shown to have bounded resource requirements.

The key to representing SDF graphs with UML activities is to recall that the SDF model is purely data-driven, in that token availability alone determines when an actor may fire. If a UML activity is developed by simply connecting a set of actions by object flows, with no control flows or control nodes, the model will represent an activity where execution is driven by the availability of data from the previous action – i.e. data-driven behavior.

This basic approach requires some refinement if activities are to represent SDF semantics, in particular with respect to token flow rules, and the actions used to represent SDF actors. Consider token flow first.

The SDF model has specific rules about token production, consumption and buffering. Specifically, when an actor fires, fixed numbers of tokens are consumed from the FIFOs associated with each input arc, and fixed numbers of tokens are produced and stored in the output arc FIFOs. The numbers of tokens produced/consumed are used to annotate the arcs as they leave/enter each actor.

In order to realize SDF graph semantics with UML 2.0 activities, multiplicities and weights must be defined. By specifying the minimum and maximum multiplicities at inputs to be equal to one another and to the corresponding input token count on the SDF graph, the associated action can be forced to execute only when that number of tokens is available. The same strategy is used with output pins, to force the production of the exact number of tokens specified by the SDF graph. The weight of the object flow from the output pin is also set equal to this number, indicating that when the source action fires, it produces a fixed number of tokens, which are immediately transferred to the input pin of the destination where they are buffered. The capacity of the input pin is specified by its upper bound [154]. Since the storage requirements of a consistent SDF graph can be determined *a priori* [16], the upper bound of the input pin can be determined.

The order in which tokens are selected to move through pins and other object nodes can be also be specified. The default 'selection behavior' is FIFO, as required by the SDF model, and so it is unnecessary to indicate this in an activity representing an SDF graph. Hence by specifying input and output pin multiplicity, object flow weights, and upper bound pin capacities as indicated above, using the default FIFO selection behavior, and by setting the pin's upper bound from the corresponding FIFO size determined from the SDF graph, then the execution semantics of SDF graphs can be represented with UML 2.0 activities.

In the majority of cases, two types of action can be used to represent actors. Computation actions transform input tokens into output tokens via the invocation of a primitive function. Primitive functions typically perform arithmetic or logical operations, and must not access or update the state of the owning object, nor interact with other objects. The representation of primitive functions is not defined by UML. Actions of this kind are used to depict atomic actors in SDF graphs, since neither may be refined within the respective models, both are triggered by the availability of input tokens, and both must execute without side-effects.

The other type of action required is the call action. Call actions are concerned with invoking behavior and receiving output values once that behavior has completed. In terms of the UML 2.0 meta-model, CallAction is an abstract class. The particular child class of interest here is CallBehaviorAction that may be used to invoke another activity, either synchronously or asynchronously. Synchronous actions of this type are used to model non-atomic SDF actors, since they are represented by SDF graphs, and SDF graphs are modeled by activities in this work.

The final elements of the SDF model that require representation in an equivalent UML 2.0 activity are fork nodes and source and sink nodes. SDF fork nodes are simply represented by fork nodes in UML 2.0, since both have the effect of duplicating the tokens arriving on the input arc on all of the output arcs [21]. Note that fork nodes are the only type of control node that appear in a UML activity that represents an SDF graph. Source and sink nodes, representing inputs to, and outputs from, an SDF graph, are modeled with parameter nodes.

On the basis of the above discussion, it may be concluded that an SDF graph can be realized as a specialized form of activity via the mapping given in Table 3.1. Such an activity will be termed an SDF activity, and a class/object whose behavior is purely represented by an SDF activity will be called an SDF class/object.

Table 3.1. Mapping between SDF Graphs and UML Activities

SDF Graphs	UML
Actors (primitive)	Computation Actions – primitive functions
Actors (hierarchical)	CallBehaviorActions
Fork node	Fork Node
Source/Sink Nodes	Parameter Nodes
Edge	Object Flow
Implicit FIFO	Pins
Token	Token
Firing Rule	Input/Output Multiplicities, Weights

3.4 Communication and Concurrency

An SDF class essentially encapsulates an algorithm, and hence instances are 'algorithm objects' which [72] indicates are common in real-time systems, and in scientific/engineering applications. Hence SDF objects are essentially equivalent to the blocks used in block diagram languages for DSP [116], and enable DSP/video processing to be described within the same system model as state-based objects. In Section 3.7 the interaction between state machines and SDF activities within objects is discussed. Here consideration is given to the way in which SDF objects communicate within UML object models, and the consequent concurrency issues that arise.

Typically SDF objects will represent different stages in a sample-processing pipeline, with the samples originating from an analogue-to-digital converter, or from an object that generates a stream of samples (the ADC and DTMF Dialer objects in the case study of Section 3.9 are examples of these two types of source). In terms of UML activity modeling, the activities of the component SDF objects are simply part of a larger activity flow, and this can be represented in an activity partition, which shows the composite activity flow, with the objects that contain the different parts of the flow superimposed in the appropriate places. However, this simply provides a view of the overall activity and its partitioning into objects, and token flows cannot be represented within object models (e.g. interactions/composite structures etc) that realize the overall activity. Such object models realize the token flows via message exchanges, since all communication in UML is message-based. This leads to object models that are low-level, and implementation-oriented.

Stream-based communication is a common and natural mechanism in many embedded systems. Indeed, abstractions of this mechanism can be found in many formalisms used for modeling embedded system behavior e.g. SDF [116], LUSTRE [83], etc. Hence stream-based communication ought to be a first-class citizen of any language that is used for embedded system specification. Unfortunately, stream-based communication is absent from UML and most other object-oriented methods (with the exception of MOOSE [136]), presumably because it can imply the breaking of information-hiding rules, which in general is undesirable. However, this is rather less important where there is a simple producer-consumer relationship between objects.

Since the purpose of this work is to develop a high level, implementation-independent notation for the specification of complete embedded systems, a new inter-object communication mechanism is proposed for UML: the *object-stream*. Object-streams define the unidirectional transmission of object or value tokens of fixed type between producer and a consumer objects, and in meta-model terms they are classifiers, meaning that they are generalizable elements. Token order is preserved within an object-stream, and so if datum i is generated by the producer, before the generation of datum j, then i will arrive at the consumer before j, for all i and j.

In a sense, object-streams represent the object flows shown in activity partitions between the activities of different objects representing the partitioning of the overall activity. However, the syntax of UML does not allow object flows to be used directly in object models.

The interface of a class whose instances receive an object-stream must declare that its objects are willing to receive a stream of a certain type, and this is done by declaring an *inlet*. An inlet is analogous to a reception, which is the mechanism by which a class indicates in its interface that it is prepared to receive signals of a particular type. In the same way that a reception may be associated with a state machine, leading to received signals being added to the state machine's event pool, an inlet is associated with an activity, via a token queue, which is an input parameter node of the activity.

A key aspect of object-stream semantics is the lack of control flow. There is no sense in which the sender executes a method in the receiver. The object-stream model simply specifies that the receiver object will provides a place for the sender to leave tokens. It is entirely up to the receiver to remove the tokens that it receives from this place, in good time.

The lack of explicit control flow has implications for concurrency. In the object model, the sending/receipt of messages can be synchronous or asynchronous [55]. Either way, activity is triggered in the receiver, and either executed on the thread of control of sender or receiver (if active – see below). In the object-stream model, there is no sense in which the writer of the object-stream directly triggers activity in the receiver. Data is simply made available in a sequence. Hence there is no notion of a transfer of control from sender to receiver. This would seem to indicate that an object that receives a stream should be active, that is, it should have its own thread of control. However, it is possible to imagine a software implementation where the receiver is a passive object (one that does not possess a thread), which is called periodically by an active object, for example, to process object-stream data that has arrived since the last call, or to execute an iteration of the SDF graph schedule. In such a case the SDF activity would represent a method that could be called, but which would suspend if insufficient data were available. Therefore, it may be concluded that classes whose instances receive streams may be active or passive, although they will often be active.

3.5 Class and Object Relationships

This section presents a discussion of the way in which typical OO concepts are interpreted when SDF graphs are utilized for the description of class behavior. Specific implementation issues relating to hardware and software are discussed in Section 3.8.

3.5.1 Dynamic Object Creation and Deletion

Conceptually, object creation poses no difficulties, in that the SDF activity is a private behavior of the class, and if the class is active, then execution can begin as soon as the contained thread is created, and data is available. In a similar fashion, object destruction is no more of a problem than system termination in traditional SDF graphs - a matter that is not normally considered in the SDF literature, since it is outside the scope of the SDF model, involving as it does, the notion of control. However, the implementation must provide some form of support to facilitate the connection of input and output streams to such a dynamically created instance, and the removal of such connections when an instance is destroyed. This is clearly a greater challenge if objects are implemented in hardware rather than in software. However, recent advances in the management of the dynamic partial reconfiguration of FPGAs may reduce the level of difficulty associated with dynamic object management in hardware [55].

3.5.2 Associations

According to [204], two classes A and B are associated if:

1. An object of class A creates an object of class B.
2. An object of class A sends a message to an object of class B.
3. An object of class A has an attribute of class B (or a collection of objects of class B).
4. An object of class A receives a message with an object of class B as an argument.

The issue of object creation has been considered in the previous section. If we extend statement (2) by replacing 'sends a message to' with 'communicates with' then this case covers object-streams, and is a logical extension to basic UML. Associations of type (3) will be dealt with below when we consider composition/aggregation. As with point (2), point (4) requires reinterpretation in the context of non-message-based communications. Since an object-stream is a classifier, it can represent a stream of objects of a particular class. Hence associations based on (4) may be interpreted in the sense that an object of class A receives a stream whose elements are of class B.

3.5.3 Inheritance

In representing class behavior via alternative models of computation, the issue of inheritance must be addressed, and the meaning of inheritance must be defined. The inheritance relationship between two classes is such that the inheriting class (or *derived* class) inherits all the characteristics (operations and attributes) of the base class whilst adding additional attributes and operations. The meaning of inheritance for classes whose behaviors are described by state machines is discussed in [193, 49], and the treatment of inheritance between classes whose behavior is described by SDF graphs is similar. Briefly, if a class X has its behavior described by a state machine, and a class Y inherits from that class, then Y's state machine may extend X's in a number of ways. New states and transitions can be added to the derived class as desired (including decomposing a state into sub-states and/or orthogonal components). Activities and actions associated with states and transitions can be added/removed/specialized for each transition/state, and the target state of transitions can be changed.

The application of these ideas to SDF classes is now considered. The SDF activity of the base class can be refined in the subclass by the modification of its actions, e.g. by replacing a primitive action with a CallBehaviorAction that invokes a new activity. However, at the higher level, the produced/consumed token counts must not be changed. Alternatively, the SDF graph of the base class can be extended in the subclass such that it forms a *subgraph* of the derived class SDF activity. Finally, both of these modifications can be applied, so that the SDF activity of the derived class can both refine the actions of the base class, and extend the base class graph.

By using the notion of clustering in SDF graphs, a relationship between the schedules of the base and derived classes can be obtained. For the sake of simplicity, this is discussed in terms of SDF graphs rather than activities. However, the discussion can be applied to SDF activities via Table 3.1.

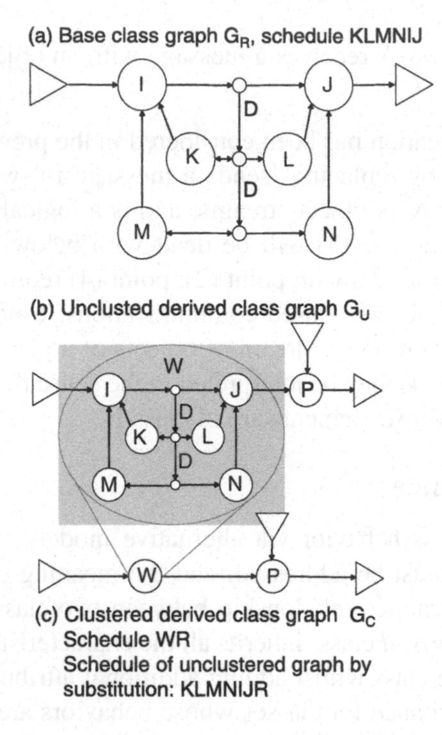

Figure 3.3. Clustering and Inheritance (SDF graph taken from [116])

Clustering is where a subgraph of an SDF graph is encapsulated into a single node to produce a modified SDF graph (see Figure 3.3). This is done when some part of an SDF graph can be treated as a subsystem [16]. In this context, if the base class is extended such that its SDF graph is a connected subgraph of the derived class SDF, then the base class subgraph can be clustered into a single node, leaving a simplified base class graph. Let the unclustered, *derived* class SDF graph be G_U, the *base* class graph be G_R with schedule S_R, and G_R be clustered into a single node Ω to produce a simplified, clustered version of G_U named G_C, with schedule S_C. Then if every occurrence of Ω in S_C is replaced with the schedule S_R of the subgraph that it represents, the resulting schedule S_U is a valid schedule for the unclustered graph G_U[16]. See Figure 3.3.

This result has significance whether the SDF graph extension is accomplished by the refinement of actors, or by the extension of the base class graph, or both. If inheritance is via actor refinement, then if the schedule of the base class SDF graph is known, along with the schedule of the graphs that refine the actors, then the schedule of the derived class can be found by substitution. Equally, if the base class SDF graph is a subgraph of the derived class SDF graph, then it can be represented by a single, clustered node, as shown in Figure 3.3(c). If the schedule of this simple SDF graph can be found then that of the full derived class SDF graph (including the expansion of the clustered node) can be found by substitution. Alternatively, the derived class schedule can be computed directly using a tool like Ptolemy [115].

3.5.4 Composition and Aggregation

Consider first the issue of composition, as this is the simpler case, since there is no difference in the lifetime of the composite and the components[1]. If a composite class is specified, which has its own behavior defined by an SDF activity, and whose component classes also have their behaviors defined by SDF activities, then what can be said about the overall behavior of the composite? In order to answer this question, the way in which the composite is defined must be specified. If it is merely by a conventional class diagram with composition 'diamonds' then it is not possible to determine the overall behavior of the composite, as there is no mechanism for specifying how the components are connected into a processing chain. See Figure 3.4.

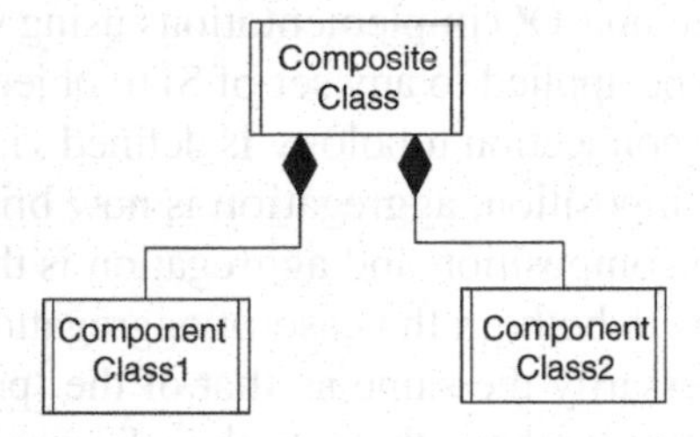

Figure 3.4. Class Diagram Showing Two Whole-Part Relationships Defined by Composition

Hence some supplementary form of diagram would be needed in this case. This is analogous to the situation with state machines discussed by [222], where sequence diagrams are used to specify the collaboration of objects that provides the behavior of the composite. However, if the components are drawn in an activity partition, which shows both the activities and instances to which they belong, then the interaction of components, and the behavior of the composite is clear. See Figure 3.5.

It is simple to determine the overall behavior of an object that is the composite of a number of SDF objects. In the general case, if the SDF activities in each

[1]Note that the word component is used in the sense of a 'part', not in the UML sense.

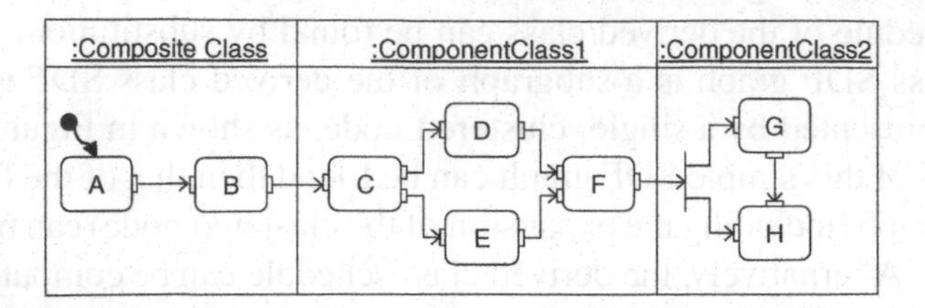

Figure 3.5. Activity Partition Equivalent to Figure 3.4

partition (object) are replaced by a single CallBehaviorAction node that invokes the relevant activity, then the overall graph typically has a very simple structure, and a high-level SDF schedule can easily be constructed, often by hand. If the schedules of each of the component graphs are known, then the schedule of the composite and its parts can be determined by substitution, as discussed in the previous section. Hence, when it comes to implementation, the behaviors of the components can combined and mapped to a single large SDF graph, which can then be implemented by standard techniques in software or hardware [116, 229]. The approach therefore allows dataflow components to be integrated into a complete object model of the system during specification, whilst supporting the synthesis of efficient, non-OO implementations using well-known methods. This approach can also be applied to any set of SDF objects that are not part of a composite, but whose connection topology is defined via an activity partition.

Having dealt with composition, aggregation is now briefly considered. One key difference between composition and aggregation is that, whilst the whole-part relationship applies to both, in the case of aggregation, the lifetime of the components is not necessarily the same as that of the 'parent'. This does not require any special treatment, above that which is discussed in this section, and in Section 3.5.2.

3.6 The Shell Model

Embedded SoC applications typically require high levels of internal parallelism, since they must typically respond concurrently to stimuli and data streams from many different sources in the environment of the system. Multiple processors, multi-tasking operating systems and application-specific hardware, which is inherently concurrent, are all employed to meet this need. Hence, in developing specification models for embedded SoCs, concurrency must play a prominent role.

The early stages of the HASoC method are concerned with use case and class/object modeling, and adopt a standard UML-based approach. After this, the model under development is organized on the basis of the major concurrent elements in the object model. These are represented by *shells*, a shell being a specialized form of active object. Shells are a generalization of *capsules* from UML-RT [194], which are themselves based on actors from ROOM [193]. A capsule is a stereotyped form of active class, with specific execution and communication semantics. The behavior of a capsule is typically represented by a single state machine, although capsules may contain sub-capsules as well as, or instead of, the state machine. Hierarchical refinement of capsules can be continued to any depth, and is depicted in capsule collaboration diagrams, which show both child capsules contained within the parent and the parent's state machine (if it has one), represented by a rounded rectangle. Capsule instances communicate via owned objects called *ports* that realize specific *protocol roles*. A protocol role defines the UML signals that are sent and received by a port. Connectors between ports in different capsules show message communications paths. The execution model is based on the transmission and reception of UML signals through ports that cause state transitions within the capsule's state machine. Computation and communication can be associated with transitions, and state entry, exit and occupancy.

A number of features from UML-RT have been included in UML 2.0. In particular, internal structure diagrams, showing the structure of a class in terms of the instances that it owns, are very similar to capsule collaborations. Ports are also included in UML 2.0, and are features of a class. When an instance of the class is created, instances of its ports are created and these are termed interaction points. If a port is specified to be a behavior port, it relays incoming information to the internal behavior of the instance of the owning class. Otherwise information will be sent to one of the objects within the internal structure of the owning instance. Behavior ports are distinguished by being connected to a rounded rectangle within the owning instance which represents its behavior. Behavior ports can be hidden within a class, in which case they are used to support communication between the internal objects and the composite's behavior.

The capsule was chosen as the basis of HASoC models for a number of reasons. The use of ports leads to the separation of communication from computation, which is recognized to be a key SoC design principle [177]. Furthermore, the capsule concurrency model is simple and clean, minimizing the possibility of interthread interference, and since most communication is by asynchronous signals, the danger of deadlock is reduced.

It is clear from the above description, however, that capsules can only represent reactive behavior, and so the capsule model has been generalized to support different forms of behavioral description. This is done by introducing the notion of a *shell*, which is an active, abstract class that has no concrete form of behavioral description, and which communicates through ports. Two specializations can be derived from the basic notion of a shell: AShells and MShells. AShells have arbitrary behavioral descriptions (e.g. in terms of code in a high-level language or hardware description language) and their ports simply encapsulate whatever form of communications is necessary to support the *ad hoc* behavioral description. AShells are concrete classes that can have direct implementations.

The behavior of an MShell is described via one or more models of computation, and the form of port communicators must be consistent with the model. MShells are abstract and are specialized into CShells (control-oriented shells), DShells (data-oriented shells), and HShells (hybrid shells). CShells are equivalent to UML-RT capsules, and have state-based behavioral descriptions and ports which support only communication by signals (signal-ports). DShells are active SDF objects that communicate via ports that transmit only object-streams (stream-ports). HShells (hybrid shells) have both control and data-oriented behavior and both signal- and stream-based ports. It should be emphasized that the behavioral descriptions exist within the context of the enclosing object, and although they may be refined according to the semantics of the prevailing MoC, the refinement does not explicitly include objects.

Signal-ports and stream-ports are represented by modifications to the basic port notation. These symbols are shown in Figure 3.6 along with examples of CShells, DShells and HShells.

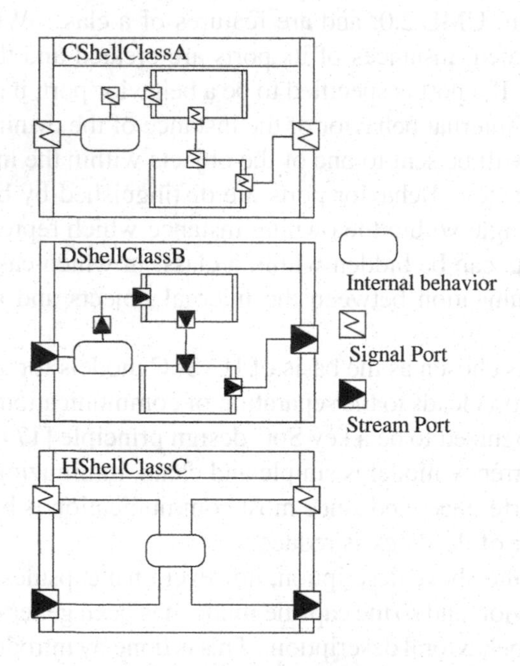

Figure 3.6. CShells, DShells and HShells

3.7 Hybrid Shells

A key issue that needs to be addressed within the notation is how the different behavioral descriptions of hybrid shells relate to each other. Consider the hybrid shell from Figure 3.6, HShellClassC. Although the internal behavior of this shell is represented by a single symbol, the fact that it is an HShell and has different types of port indicates that it has a state machine, and one or more SDF activities. In the general case, the state machine and the SDF activities will interact, since if the descriptions are independent, then combining them within a single class is inappropriate on the grounds of poor cohesion. Hence in this formulation of the hybrid shell model, an SDF activity is simply represented as a form of in-state activity, and so the state machine controls the execution of the SDF activity. This is a more conservative approach than the one proposed by [68] which permits the arbitrary nesting of SDF graphs within states, and vice versa. The more constrained approach was adopted because in this model, classes and objects are the major structural elements for system description, and although they may contain arbitrarily complex behavior, consideration of good OO design would typically militate against objects with very high degrees of internal complexity. Hence the general approach of [68] is not adopted, although some notions from that work are utilized.

In terms of UML state machines, a behavior that is triggered upon entry to a state, and which executes whilst the machine remains in that state is known as a *do-activity*. The UML semantics of do-activities are defined in [154] as follows:

1. Do-activities start upon entry to a state.
2. If a do-activity is executing when an event arrives triggering a transition from the enclosing state, then the do-activity is aborted.
3. If the do-activity completes whilst the state machine resides in the enclosing state, then a completion transition is triggered.

If an SDF activity is associated with a do-activity, then rule (3) is irrelevant since the SDF model only deals with semi-infinite data streams. Rules (1) and (2) however, require modification in the light of the SDF model. In order to explain this, consider the following state machine that represents the state-based behavior of an HShell HShellClassC.

The role of the state machine shown in Figure 3.7 is to initiate, pause, resume and terminate the SDF behavior S. The SDF is represented by a do-activity in state Active. To provide control, a number of actions are defined that enable the state machine to interact with the SDF activity. These actions, which all apply to a named SDF graph (S in the example) are associated with transitions or state entry/exit in the normal UML manner, and are all logically transparent to the SDF model. The actions are:

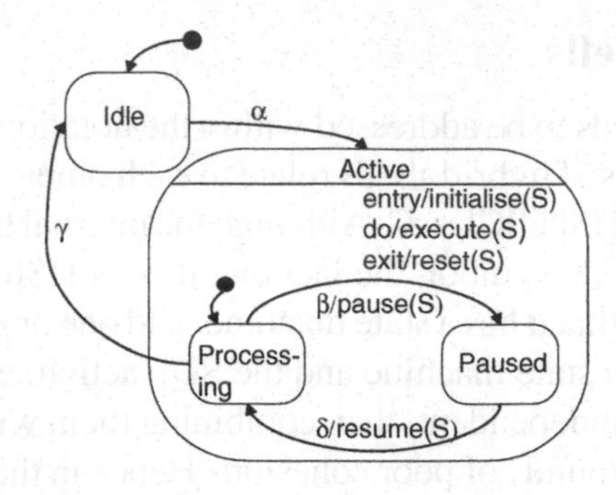

Figure 3.7. Example State Machine

- Initialize(S): place initial values on the SDF arcs.
- Execute(S): executes the SDF.
- Reset(S): clears values from the SDF arcs.
- Pause(S): stores the current values on the SDF arcs and then executes a reset.
- Resume(S): restores values stored by the previous pause operation. The result of executing resume when no previous pause has been executed is undefined.

Note that the execute action is not strictly necessary, as once an SDF activity has been initialized or resumed its execution is driven by the availability of input data through the enclosing object's stream-ports. It is included for documentation purposes. If any of these actions are associated with states or transitions, they override rule (2) above. This requires a change to the existing UML semantics. For example an event that causes a transition out of a state with a do-activity may either abort or pause the do-activity, depending on the action associated with the event.

The above operations are chosen in such a way as to enable the SDF to be controlled (i.e. started, paused, resumed etc) without the need to modify the SDF model in any way at all. The operations simply achieve their effect by removing or restoring tokens, relying on token presence/absence to control the SDF operation. It is clear that an implementation would have significant latitude with respect to how the same effects are achieved.

The above considers the control of an SDF activity by a state machine. Influence may also be exerted in the opposite direction: the activity may force the state machine to change state, via the addition of SendSignalActions to the SDF activity. This is accomplished by adding a CallBehaviorAction to the SDF activity, whose role is to evaluate a data dependent condition and if appropriate, to generate a SendSignalAction [154] to force the state machine to change state. This additional action has no effect on the overall behavior of the SDF activity, and so does not affect the SDF activity schedule.

3.8 Mapping Shell Models to Software and Hardware

Straightforward mappings of shell models into OO languages can be achieved. However, since there are many well-known implementations of state machines and dataflow graphs, it is also possible to target lower level, non-OO implementations. For example a software CShell can be mapped to an OS thread with a message queue, and in very primitive execution environments, threads may be dispensed with altogether, and a cyclic executive structure can be used [179].

The implementation of DShells is, to some extent, bound up with their mapping to execution engines. For example, DShells would typically be used to represent different filtering stages in a DSP chain with, for example, the output of one stage providing the input to the next. In such cases, the most efficient implementation would be obtained by exploiting the composition relationship mentioned in Section 3.5 and composing the SDF graphs of all DShells allocated to a common execution engine into a single SDF graph, and then applying standard SDF graph implementation techniques [16, 229].

Implementations of HShells must consider the interaction model, discussed in Section 3.7. Key issues include the concurrent, or pseudo-concurrent, execution of the state machine and SDF graph, and how the SDF graph is paused or reset upon state exit/transition/entry. In a single threaded software implementation, once a state is entered in which the SDF graph executes, the thread must be used for this purpose. However, to avoid state machine starvation, checks for event arrival must be made from within the SDF graph schedule. Such checks are still necessary even if the SDF graph runs on its own thread in a multi-threaded implementation, unless the OS supports some form of asynchronous signaling.

If a hardware implementation is chosen for a shell (or set of shells), then there are several issues that must be addressed. These include the integration of the shell with the underlying platform, the level at which the shells are to be represented, and communication between the shell and the rest of the platform.

In terms of HASoC, a shell that is chosen for hardware implementation is both part of the applications model and part of the underlying platform. The system platform is initially modeled as a UML deployment diagram, from which a shell model of the platform is synthesized [78]. This is known as the hardware architecture model (HAM) and provides a framework for simulation and synthesis, by supporting the integration of IP blocks, or the development of hardware modules from scratch.

Hardware modeling in HASoC is conducted either at the transaction level or register transfer level. Transaction level modeling is used to support rapid simulation early in development and consequently uses high level behavioral models, and intermodule communications based on interface functions in channels that encapsulate the low level details of data transfer. SystemC has proved to be popular for transaction level modeling [92], and a mapping between the HAM and SystemC has been established [73], and a prototype tool that generates SystemC from simple shell models has been developed [2]. In such models, high level representations of state machines and SDF graphs can be used, since the aim is to support simulation. Communication by simple object streams and events is realised by calls to channel interface functions, necessitating the addition of detail to the ports of hardware shells.

In terms of developing models that are suitable for synthesis, the behavior of the shell must be represented at the register transfer level, and inter-module communication must be refined to this level of detail. In terms of internal behavior, techniques have been developed for the synthesis of register transfer level descriptions of SDF graphs and statecharts (upon which UML state machines are based). See, for example, [16, 223]. Hence attention is focused mainly on the implementation of the interaction model for HShells. However, it is necessary to explain that the SDF graph is represented as a set of execution units (EXUs), which can implement either actors or precedence graph firings, interconnected by token storage. The firing of the execution units is controlled by the SDF graph scheduler, which implements the firing schedule of the graph. See Figure 3.8. The state machine actions that control the SDF graph (e.g. see Figure 3.7) are mapped directly to control signals that are input to the SDF graph scheduler (Figure 3.8). These are distributed to the EXUs, along with the basic control signals needed to control the EXUs.

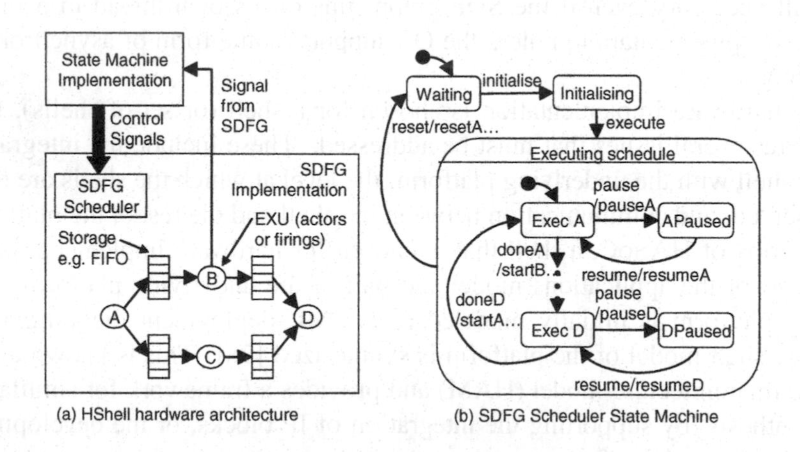

Figure 3.8. Hardware Implementation of an Hshell

In addition to the representation of shell behaviour at the register transfer level, intermodule communication must also be represented at this level. A procedure for refining communications in a HASoC HAM from transaction level to register transfer level is presented in [73], which is based on a procedure discussed in [162]. In the initial stage of communication refinement, the ports of a hardware shell are replaced by an adapter port that translates between the transaction level interface and the bus protocol, enabling an RTL simulation of communication to be carried out, whilst using transaction level models of behavior. When a register transfer level description of shell behavior is available, the adapter port is replaced by one with the same external interface, but whose internal interface is now at register transfer level.

3.9 Case Study: A Simple Modem

A small case study will now be presented. The system under consideration is a simple modem, which is intended to be a subsystem of an embedded SoC. A number of assumptions and simplifications are made in the interests of brevity. The modem and the 'main' system communicate through shared memory, and so there is no need to model a serial interface, AT commands and an AT command parser. Moreover, it is assumed that the modem operates at a single line-speed, and so there is no negotiation between sending and receiving modems, once a connection is established.

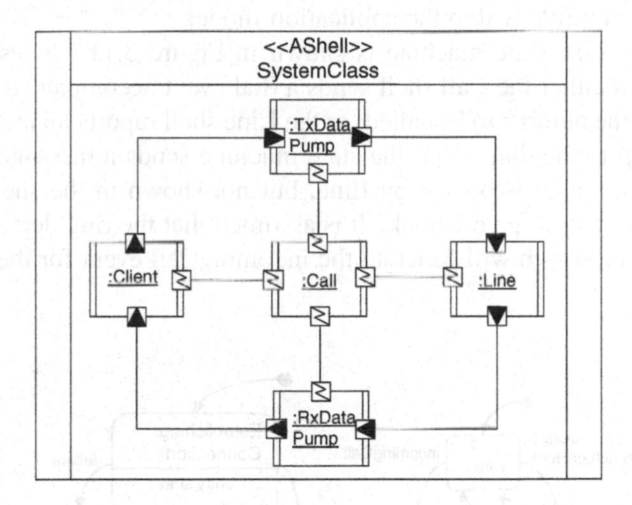

Figure 3.9. Modem System Shell Diagram

A shell model of the system is shown in Figure 3.9. The Client shell represents the part of the SoC that uses the modem to transmit messages. The Call shell is a high level object that tracks the status of a call, and which mediates between the Client and the Line. The Line shell encapsulates all the details of access to the physical communications channel. The DataPump shells are responsible for most of the signal processing operations that must be performed by the system e.g. modulation/demodulation/filtering etc etc. In terms of the shell categorization given in Section 3.6 the Client is an AShell, Call is a CShell, and the Line and the Data Pumps are HShells.

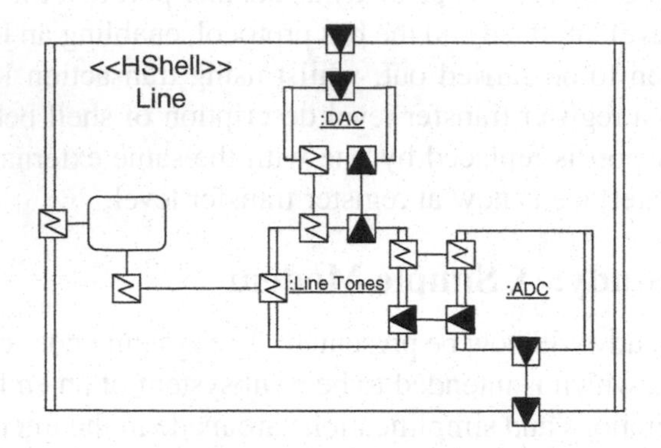

Figure 3.10. The Line Shell

The refinement of the Line shell is given in Figure 3.10. In terms of a specification model, the ADC and DAC represent hardware components that play a significant role within the application model.

Part of the Line state machine is shown in Figure 3.11. It resides in the Idle state until either the Call shell sends a dial event accompanied by a string representing the number to be called, or the Line shell reports an incoming call. In responding to the dial event, the state machine sends a message to passive object switch, which is owned by Line, but not shown in the shell diagram, causing the system to go off-hook. It is assumed that the ring detection is part of the platform, which will generate the incomingCall event for the Line shell state machine.

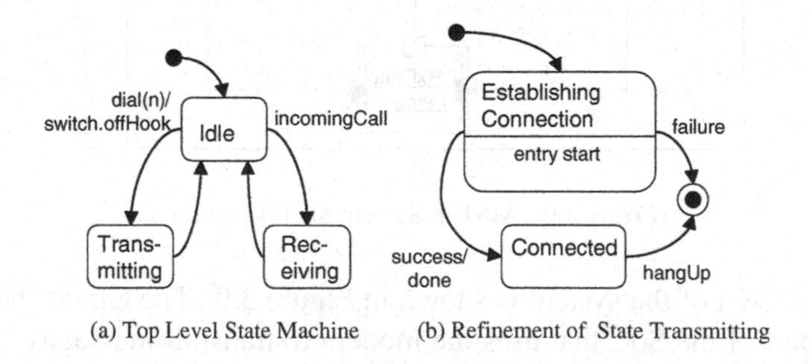

Figure 3.11. Part of the Line State Machine

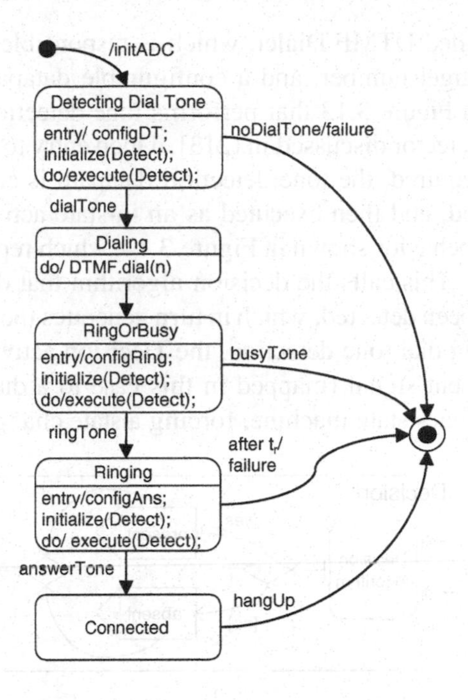

Figure 3.12. LineTones State Machine

The refinement of the Transmitting state is shown in Figure 3.11(b). The main task here is to initiate, via the start entry action, the connection sequence that is performed by the shell LineTones in Figure 3.10. If LineTones is successful in making a connection, it sends event success to the Line state machine, which in turn sends a done event to the Call shell. This switches the TxDataPump on and instructs the Client to send data to it. If a connection cannot be established then a failure event is sent to Line.

The LineTones state machine (Figure 3.12) implements the sequence of operations required to establish a connection, specifically dialing tone detection, DTMF dialing, ring/busy and answer tone detection. The LineTones shell owns

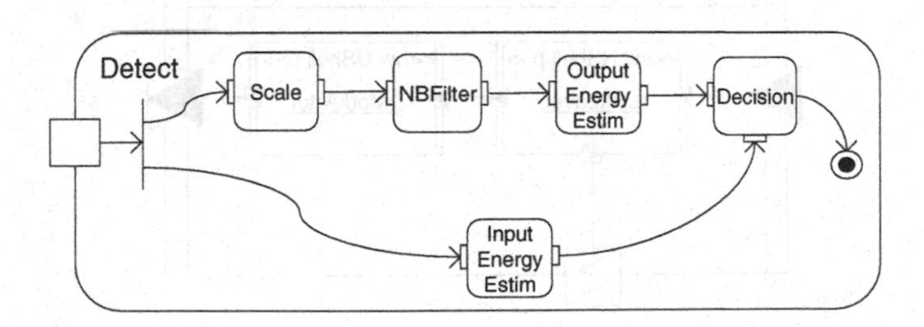

Figure 3.13. SDF Activity

a passive object called DTMF Dialer, which is responsible for generating the tones to dial the target number, and a configurable datapath, represented by the SDF activity in Figure 3.13 that performs tone detection. This activity is based on the tone detector discussed in [213]. Upon entry to each state in which tone detection is required, the tone detection datapath is configured, the SDF activity is initialized, and then executed as an in-state activity. The Decision action invokes the behavior shown in Figure 3.14, which represents the generic detection behavior. This calls the decision algorithm that determines whether or not the tone has been detected, which in turn generates the appropriate signal. For example, during dial tone detection, the Decision activity will send either a present or an absent signal (mapped in this case to a dialTone/noDialTone event) to the LineTones state machine, forcing a state change.

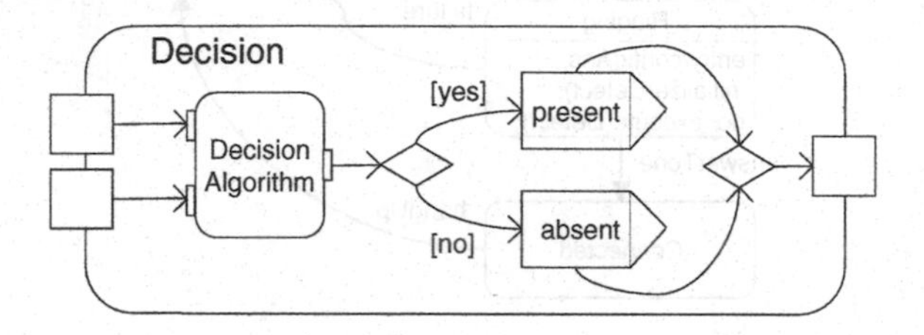

Figure 3.14. Decision Activity

3.9.1 Commitment and the Platform

In HASoC, shells are committed to software or hardware implementation, typically on the basis of non-functional requirements. In this section a small part of the modem model will be committed in order to illustrate the process.

Figure 3.15 shows the decomposition of the TxDataPump shell into a LinkLayer HShell that is responsible for link management, and a Modulator DShell that contains a chain of DShells that perform all the signal processing necessary to produce a modulated sample stream that is sent to the DAC.

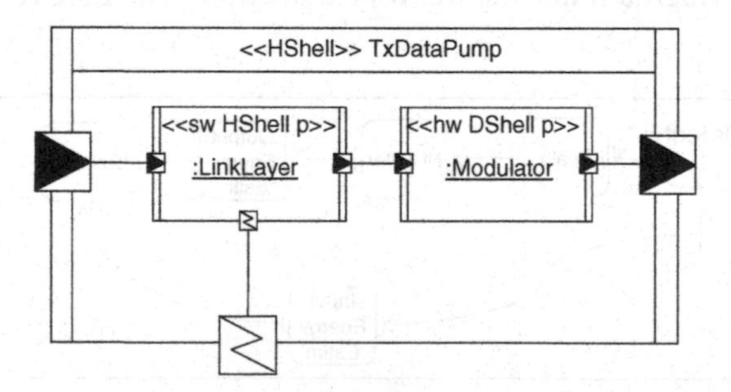

Figure 3.15. Decomposition of TxDataPump

It is assumed that, on the basis of performance estimates, the decision has been taken to implement the LinkLayer in software and the Modulator in hardware. The identifier 'p' in the stereotype indicates that the LinkLayer shell will run on processor p which is part of the platform (see below), and that the Modulator will be interfaced to p, and share memory with it.

As indicated in Section 3.8, application-specific hardware shells are also part of the system platform, which is represented by the HAM. In this simple example, it is assumed that the initial system platform consists of a single processor, SRAM, ROM, a DAC and an ADC, all connected via a single bus. The HAM, including the Modulator is shown in Figure 3.16. A transaction level SystemC model can be developed within this framework, as indicated in Section 3.8, provided that appropriate SystemC models of the components exist, or time is available to develop them. On the basis of data gathered from simulation runs, it may be deemed necessary to reorganize the platform, for example by introducing a more elaborate bus structure, additional memories etc.

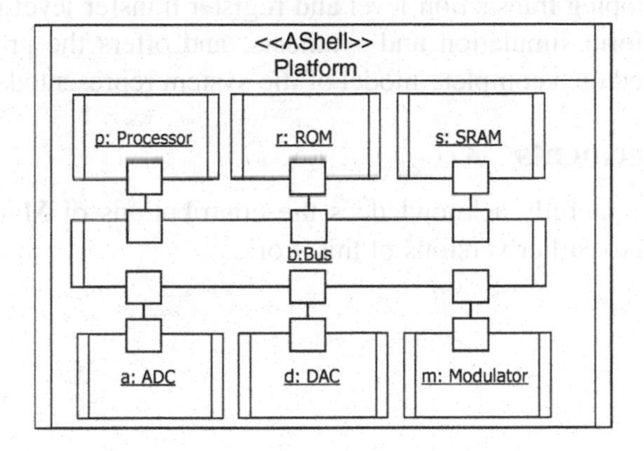

Figure 3.16. Platform Model

3.10 Conclusions

This chapter has introduced a formalism to support the specification and modeling of complex embedded SoCs. The approach is based on the HASoC design method and UML 2.0, and a key aspect is the integration of control-oriented (state machine) and dataflow (SDF) models of computation. It allows embedded SoCs to be modeled uniformly in terms of a specialized form of object, known as a shell, and for those shells to exhibit different forms of behavior. Some shells will be reactive, and their behavior is described by state machines. Others will exhibit dataflow behavior, which is represented internally by SDF activities. Finally some shells will display both types of behavior, requiring both types of internal behavioral representation. In such cases, a simple interaction model has been defined, enabling the different types of behavior to communicate. The overall approach requires a small number of minor modifications to UML 2.0, specifically the addition of object-streams and the associated constructs, and small changes to the semantics of do-activities, enabling them to be optionally resumed upon state re-entry.

Having a complete representation of system behavior enables the specification to be evaluated against the requirements of the whole system. This does not, however, prevent the different aspects of a system's behavior being considered in isolation if necessary. For example, model checking can be applied to the reactive parts of the system to establish the presence of desirable properties and the absence of undesirable behavior [212].

Shell models can be realized in software or hardware, and a range of implementation options exist, some of which can yield very efficient implementations based on well-known techniques. The HASoC method supports the integration of hardware shells with the underlying system platform, and provides procedures for developing transaction level and register transfer level models. This facilitates platform simulation and synthesis, and offers the prospect of co-simulation based on a complete model of the system represented in UML 2.0.

Acknowledgements

The author gratefully acknowledges the contributions of Martyn Edwards and Salah Essa to earlier versions of this work.

Chapter 4

A Generic Model Execution Platform for the Design of Hardware and Software

Tim Schattkowsky, Wolfgang Mueller, Achim Rettberg

Paderborn University/C-LAB
Paderborn, Germany

Abstract This chapter presents the concepts of our Model Execution Platform (MEP). The MEP is an approach to executable UML for the design of hardware and software systems covering Class, State Machine, and Activity Diagrams. We present how the MEP is employed for Handel-C code generation and briefly sketch the concepts of a MEP based UML virtual machine.

4.1 Introduction

In electronic and embedded systems design we can currently identify a number of gaps in moving from specification to implementation.

In embedded systems design most approaches employ platform specific code generation. Various code generator targets are available covering various microcontrollers such as the C166 and C8085 families as well as different Real Time Operating Systems (RTOSs) such as OSEK [128]. Many efforts have been made to investigate retargetable compilers that can be tailored to different hardware platforms [119].

In electronic systems design the platform based approach has become quite popular. Platform based chip design encompasses the use of SoC platforms and the integration and reuse of IPs [181, 34].

Most recently, UML and the idea of MDA (Model Driven Architecture) [147] has become well recognized in the domain of embedded software and hardware systems. MDA is based on the idea of platform-independent development with Platform-Independent Models (PIMs). PIMs have to be mapped to Platform-Specific models (PSMs), which are used for the derivation of the actual implementation. In that context UML has an important role for system

G. Martin and W. Müller (eds.), UML for SOC Design, 63–88.

documentation and specification. The notion of Executable UML is of major interest here as it enables a UML based PSM to become executable.

This chapter presents a model-based approach for hardware and embedded software design to bridge the gap from specification to implementation based on the notion of executable UML. For that we introduce the concepts of our Model Execution Platform (MEP). The MEP is based on a UML 2.0 subset with precise behavioral semantics. This subset covers Class, State Machine, and Activity Diagrams. It provides abstract concepts which can be employed for the hardware code generation as well as for embedded software. Based on the MEP, we introduce concepts for true object-oriented hardware design starting from Class Diagrams arriving at a synthesizable hardware description language such as Handel-C. Additionally we outline how to apply the same concepts for the efficient execution of binary encoded UML specification on a virtual machine.

The remainder of this chapter is structured as follows. The next section discusses related works. Section 4.3 introduces the basic MEP concepts. Section 4.4 employs the MEP concepts for Handel-C code generation. Section 4.5 briefly sketches how the same concepts are applied for the implementation of a UML virtual machine before Section 4.6 finally closes with a conclusion.

4.2 Related Works

The MEP is a model-based approach for the design of hardware and software systems. Related works can be mainly identified in the areas of abstract execution platforms for FPGAs, methodologies for hardware/software codesign, and the generation of hardware/system description languages from UML.

In the field of abstract hardware execution platforms, the Hardware Virtual Machine project [81] targets at the specification of an abstract FPGA in order to overcome the problem of incompatible bit files. Designs for such an abstract FPGA are automatically transformed into an FPGA bitstream file. This transformation is based on an automatic assembly from small fragments, which are further subject to place and route.

Lange and Kebschull introduce the idea of a virtual machine for abstract hardware implementations running on specific types of FPGAs [111]. Their approach is based on the execution of byte code, which essentially contains a binary encoded register transfer level description. The byte code is composed of blocks of instructions that are scheduled into multiple equal execution units within the actual virtual machine implementation. This virtual machine implementation is specific for a particular FPGA and may vary in the number of execution units. Their byte code describes low level hardware designs comparable to a direct FPGA implementation and does not support high level control constructs.

For SoC and hardware/software codesign, several methodologies start from platform-independent diagrams or C based specifications. One example is the OCTOPUS approach for embedded systems design [5]. It is based on OOA and already follows the idea of diagrammatic platform-independent specification. OCTOPUS covers hw/sw partitioning and embedded software design. However, OCTOPUS just introduces general concepts without complete tool support for specification and code generation.

The SpecC methodology is based on the SpecC language, a concurrent C extension [67]. The SpecC specification is the starting point for an architectural exploration and a communication synthesis, which gives the input for further software and hardware synthesis. The methodology comes with the SoC Environment (SCE) from UC Irvine, an advanced tool set for hw/sw partitioning and profiling.

COSYMA [90] denotes a methodology and a toolset for hw/sw codesign and cosynthesis. The design starts with a specification written in C_x. $C_x$is a C extension by processes and interprocess communication. After compilation and profiling, the design is partitioned into hardware and software, which are subject to further synthesis.

The IMEC SoC++ design flow [123] is a C++ based methodology for hw/sw codesign. It starts with a specification in concurrent C++ and covers the exploration of storage and data transfer management in addition to task scheduling and a Pareto-based power and performance analysis. The design can start with UML for system requirements and architecture. Here UML has more the role of an optional graphical frontend for system specification and is not an integral part of the design flow.

In the context of the OMG Model Driven Architecture the notion of platform-independent design and executable UML became popular for retargetable software generation. Most approaches are mainly based on Class and State Machine Diagrams (resp., StateCharts) such as xUML [169] and xtUML [131].

The concept of xUML is based on the notion of an Action Specification Language (ASL), which defines the semantics of basic actions for code generation [148]. However, for creating an executable model xUML still relies on a programming language specific code generation (e.g., for Ada, C, C++).

A similar approach is taken by xtUML, which defines an executable and translatable UML subset for embedded real time systems. The modeling tools integrate abstract, macro-like constructs, which are easily retargetable to the various C dialects of different microcontroller platforms.

Implementations of xUML and xtUML can be found in iUML (Kennedy Carter) and Bridgepoint (Accelerated Technology), respectively. Comparable approaches are taken by Real-Time Studio (ARTiSAN) and the Ameos tool suite (Aonix).

Most recently, we can find several investigations on code generation for hardware/system description language from UML like SystemC and VHDL [139]. Most of them are in more details in the other chapters of this book.

In this chapter we introduce the Model Execution Platform (MEP) which defines an abstract model-based platform for executable UML in the context of embedded systems and SoC design. The MEP is generic in the sense that it is applicable as a model-based approach for hardware and embedded software systems. In this chapter, we focus on Handel-C code generation for FPGA synthesis and only sketch the application for embedded software. Details of the MEP based virtual machine for embedded systems can be found in [185]. In contrast to other approaches, we apply a combination of State Machines and Activities for behavioral specification based on a strict object-oriented methodology. Though this chapter just presents the Handel-C code generation for FPGA synthesis, our approach is not limited to Handel-C and can be easily adapted to other system or hardware description languages like VHDL or System-C.

4.3 The Model Execution Platform (MEP)

The UML is a general purpose modeling language. Since UML models are not necessarily executable, the notion of Executable UML and UML model compilers for transforming executable UML models into implementations are of particular interest. To arrive at executable models it is necessary to tailor UML to such an application. For this a well defined UML subset for the creation of a composite UML model consisting of different views and diagrams must be identified along with its execution semantics.

The foundation of our approach is the definition of a generic Model Execution Platform (MEP) as given in the remainder this section. The MEP introduces syntax and semantics for complete system specification based on a well defined UML subset with precise execution semantics. The subset is based on the definition of a Class Diagram with classes, their properties, and operations in combination with State Machines, Activities, and Actions.

Inherently different techniques may be applied to implement the execution semantics of our MEP. In Section 4.4 we discuss the implementation through automatically generated synthesizable hardware descriptions. In Section 4.5, for comparison, we sketch the concepts of an implementation through a virtual machine (VM) (see also [185]).

4.3.1 Structure

As our approach is fully object-oriented, UML Class Diagrams provide the basis for our structural specifications where Associations[1] reflect the existing

[1] We capitalize references to classes of the UML standard metamodel.

attribute type information and Generalizations define subclasses and implemented Interfaces.

The object model of our MEP UML subset is strictly derived from the UML 2.0 metamodel (see Fig. 4.1). A *Class* supports Operations and Properties, which can be static or non-static Features. Single inheritance and the use of multiple interfaces are included. An *Operation* may have in, out, inout, and return *Parameters* with an optional default *ValueSpecification.* Furthermore, an Operation may raise typed exceptions.

In the UML 2.0 metamodel the actual method for an Operation is defined by an instance of a *Behavior* subclass. The MEP employs a StateMachine for this purpose which is outlined in more details in Subsection 4.3.2. If no Behavior is given, the Operation is implicitly defined as abstract.

Instance and class variables are declared through non-static and static *Properties,* which represent the attributes of a Class. A Property is defined by its name, type, and visibility (public/private/protected). Unlike arrays, we do not support collections. However, they can be easily implemented by run time classes.

4.3.2 Behavior

The UML offers a rich set of concepts for behavioral modeling. *Actions* represent single computational steps and are composed to Activities. An *Activity* provides means for modeling flow oriented behavior based on Petri Net-like token semantics. Activities are composed of subActivities or Actions, where the latter are the fundamental behavioral elements of the UML. An *Activity Diagram* graphically represents such an Activity.

UML StateMachines enable state-oriented modeling. *StateMachines* are based on the concept of hierarchical finite state machines. StateMachines invoke Activities when executing States or Transitions as Entry, Do, Exit, or Effect Activities. A StateMachine is represented by a *StateMachine Diagram.*

An *Interaction* describes behavior in terms of partially ordered messages between objects and is comparable to a SDL message sequence chart. An Interaction can be represented by a *Sequence Diagram.*

Actions, Activities, StateMachines, and Interactions are subclasses of Behavior in the UML metamodel and can describe the method of an Operation. In our MEP approach we use a StateMachine to describe the Behavior of an Operation. The States in a StateMachine can contain Activities, which finally contain Actions. Since Actions can invoke other StateMachines, this provides the support for arbitrarily combined data, control, and state oriented modeling for each Operation.

Actions as Basic Elements. Actions are the fundamental behavioral units in the UML. Each Action is considered as one computational step. For executable

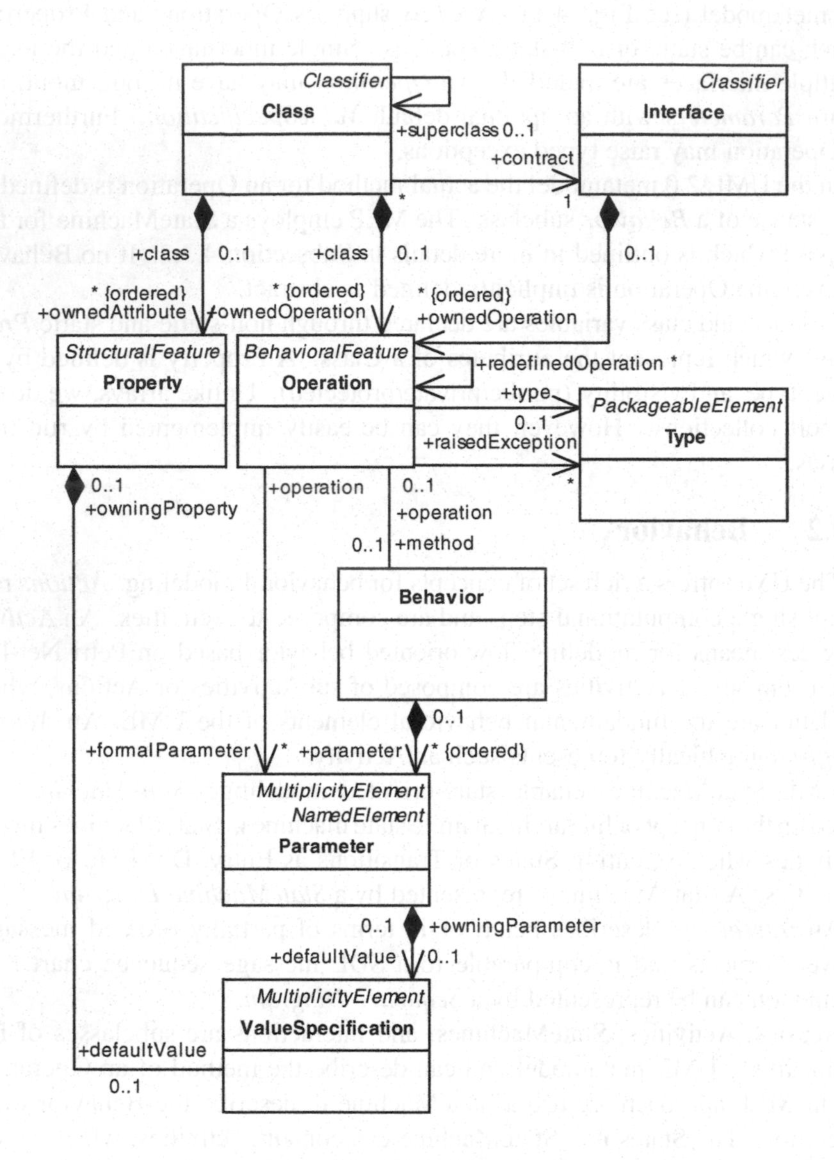

Figure 4.1. MEP Object Model

UML, Actions and the owning Activities have to form a computationally complete language. Our approach utilizes most of the Actions introduced by the UML 2.0 [157]. In cases where the UML provides no graphical notation, we employ a JAVA-like textual syntax for Action specification.

Using Activities for Control and Data Flows. StateMachines can hardly express complex behavior with nested conditional control flows, loops, and complex data structures. A typical example is the specification of the simple Bubble Sort algorithm, where no useful states can be identified. In Activities, integrated data and control token flows address the needs for combined data and control oriented modeling much better. We thus employ such Activities through their application in StateMachines.

The MEP supports the essential elements of UML 2.0 Activities such as Initial, Final, Action, Fork, Join, Decision, and Merge nodes. However, for efficient code generation we have introduced wellformedness rules for their composition. As an example we require a closing Join for each Fork element. Furthermore, as UML 2.0 introduces the definition of input and output Pins for an Action, we apply them to represent the signature of individual ActivityNodes.

Defining Operations through StateMachines. Many approaches employ StateMachines to describe the behavior of a whole Class at state level with Transitions triggered by Operation calls. Other applications are fundamentally different. They apply transitions to react to a set of heterogeneous signals, without object-oriented semantics. Thus some developers implement StateCharts through a single Operation like several reference implementations for protocols do. The definition of Operations through StateMachines enables the integration of additional model elements such as Exceptions and timeouts. Additionally it enables the dynamic composition of StateMachines through nested operation calls and also avoids the artificial creation of an individual class for each StateMachine, which appears to be more intuitive for the specification of complex systems. Therefore we follow that approach and define Operations through StateMachines. It is important to note here that a trivial StateMachine (i.e., a single State with a single Activity) is equivalent to the direct application of an Activity for the definition of an Operation.

A MEP *StateMachine* consists of simple and composite states (see Fig. 4.2). The latter can embed another StateMachine. Initial *Pseudostates* are the only supported PseudostateKind. Concurrent States are intentionally not supported. Alternatively, concurrency can be implemented through asynchronous Operation calls. This enables the dynamic creation of concurrently executing StateMachines while maintaining the UML 'run to completion' semantics.

Both composite and simple States may have Activities defining their behavior on entry, exit, and during execution. Transitions usually have explicit Triggers.

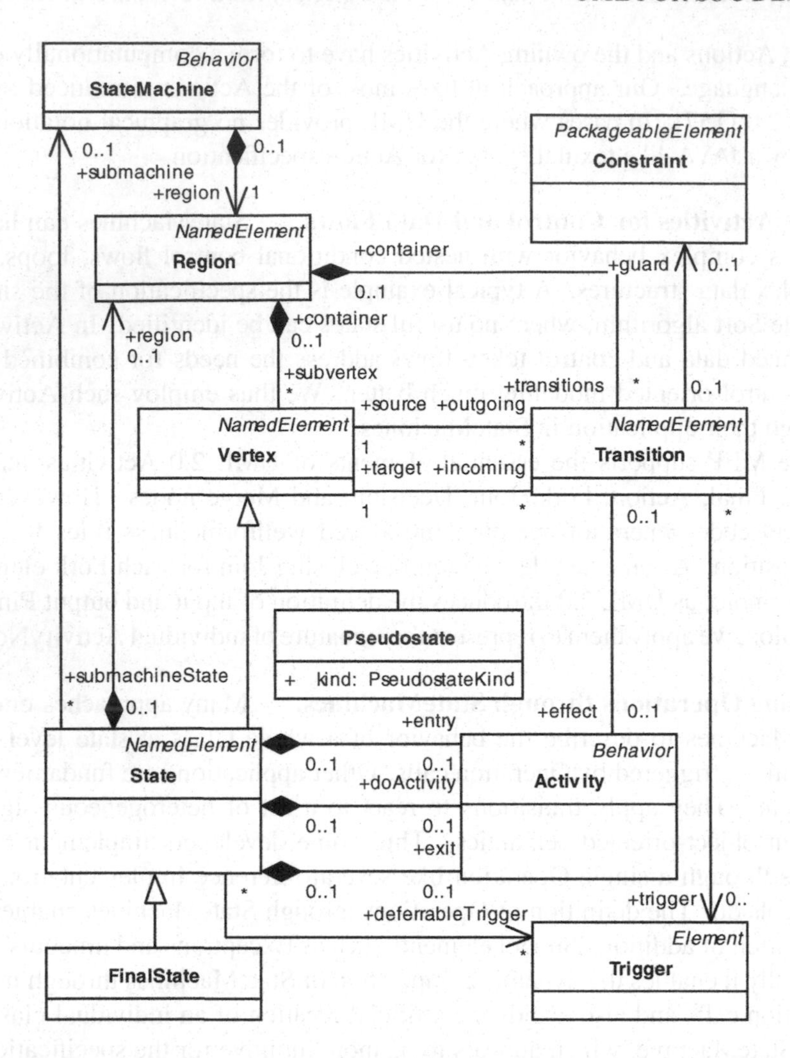

Figure 4.2. MEP State Machine Model

In addition, one unguarded completion transition without a Trigger may exist for each State. As given in Fig. 4.3, Transitions may have different *Triggers*: the occurrence of a timeout or hardware Interrupt, a software *Exception* (e.g., division by zero), or an explicit Trigger (*ImmediateTrigger*). ImmediateTriggers are instantly processed and cause the current State to be exited immediately. An Operation completes when the corresponding StateMachine terminates by reaching a final state.

The MEP supports timeouts, interrupts, exceptions, explicit events, and the completion event for the embedded Activities. Exceptions are raised from StateMachines by taking them as the effect of a Transition to a final State

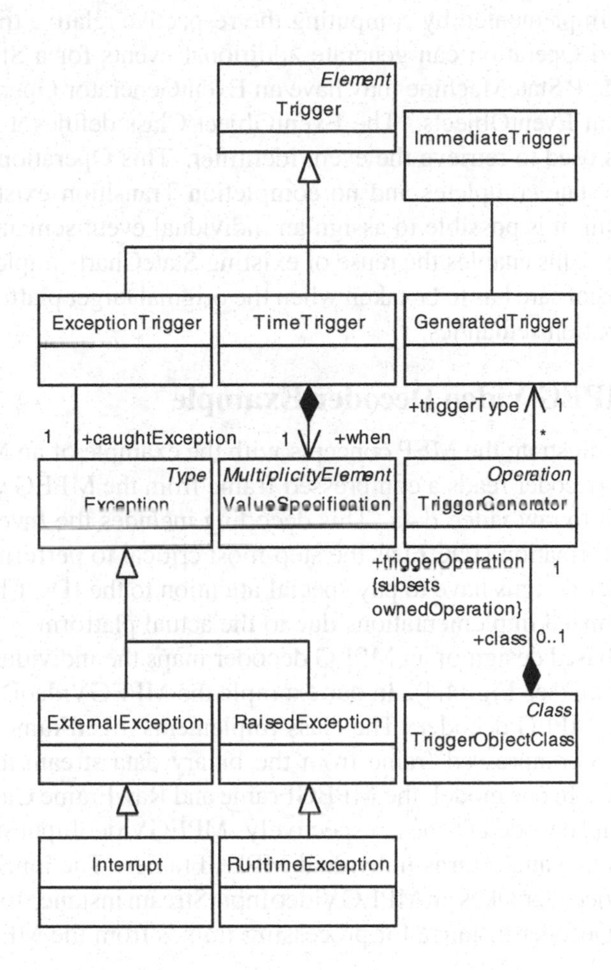

Figure 4.3. MEP Trigger Types

at the given StateMachine level. Generally, all Exceptions thrown in a sub-StateMachine have to be caught in the containing StateMachine by defining Transitions from the containing State, which is triggered by the Exception. Transitions may raise the same Exception again to propagate it in the StateMachine hierarchy. Timeouts are implemented by a TimeTrigger with an Integer value and a relative physical time unit as a parameter. Absolute TimeTriggers can be easily implemented by computing the respective relative timeouts.

A dedicated Operation can generate additional events for a StateMachine, where each MEP StateMachine may have an EventGenerator Operation to produce additional EventObjects. The EventObject Class defines a member operation that is used to retrieve the event identifier. This Operation is executed once when a State completes and no completion Transition exists. Through this mechanism, it is possible to assign an individual event semantics for each StateMachine. This enables the reuse of existing StateCharts implementations. However, special care has to be taken when the original target platform enforces different execution semantics.

4.3.3 MPEG Video Decoder Example

We briefly illustrate the MEP concepts with the example of an MPEG video decoder. The decoder reads a compressed frame from the MPEG video stream and decodes it to raw video data. This decoding includes the Inverse Discrete Cosine Transformation (IDCT) as the step most critical to performance. Thus MPEG decoder designs have to pay special attention to the IDCT by providing different optimized implementations due to the actual platform.

The MEP based design of an MPEG decoder maps the individual entities to a Class Diagram (see Fig. 4.4). In our example the MPEGVideoDecoder class represents the MPEG decoder. The class implements a getFrame() Operation that retrieves a compressed frame from the binary data stream and returns a decoded frame. In our model, the MPEGFrame and RawFrame Class represent the encoded and decoded frames, respectively. MPEGVideoInputStream parses the MPEG stream and returns individual MPEGFrames. The initStream() Operation of the decoder takes an MPEGVideoInputStream instance to initialize an MPEGVideoDecoder instance for processing frames from the MPEGVideoInputStream.

To decode a frame, the private decodeFrame() operation uses a dedicated implementation of the IDCT Interface. Through this Interface different IDCT implementations can be employed depending on the characteristics of the individual hardware or software platform.

We will outline the getFrame() Operation here in more detail to illustrate the design of behavioral aspects. The getFrame() Operation is specificed by a StateMachine (see Fig. 4.5) with a single State, which executes the Process-

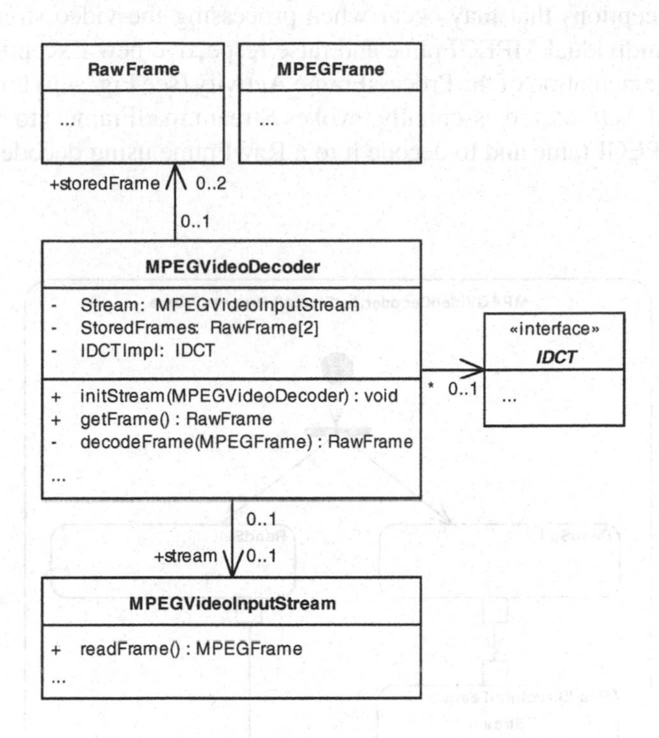

Figure 4.4. Simplified MPEG Video Decoder Class Diagram

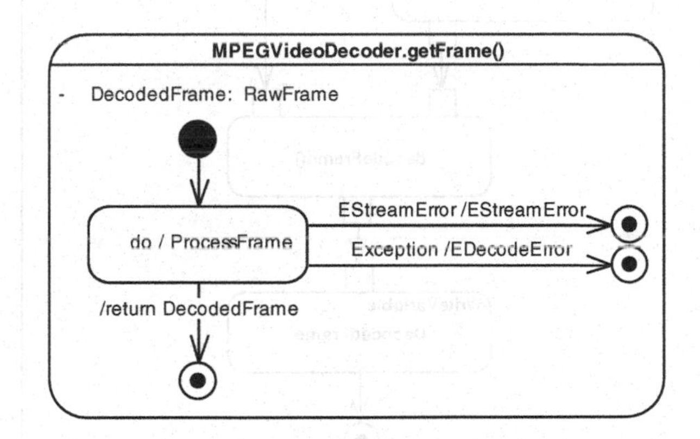

Figure 4.5. MPEGVideoDecoder.getFrame() State Machine Diagram

Frame Activity. Note here that the StateMachine is not trivial, as it is employed to catch Exceptions that may occur when processing the video stream or decoding the individual MPEGFrame and raise respective new Exceptions.

The implementation of the ProcessFrame Activity (see Fig. 4.6) is composed from several Actions and essentially invokes Stream.readFrame() to retrieve an encoded MPEGFrame and to decode it to a RawFrame using decodeFrame().

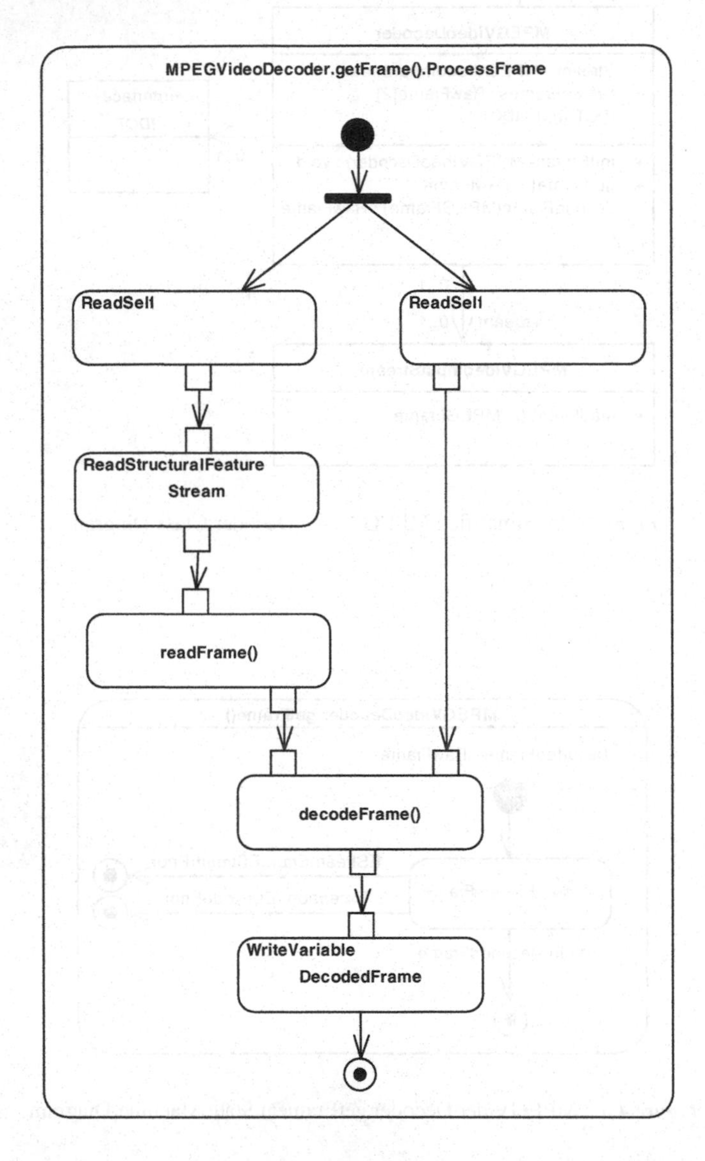

Figure 4.6. MPEGVideoDecoder.getFrame().ProcessFrame Activity Diagram

Based on the Class Diagram we can decide on a first partition into hardware and software components. As the IDCT is known as performance-critical, it is a candidate for implementation on a DSP or dedicated hardware. The example in the next section outlines how to implement the IDCT as a butterfly network and generate Handel-C code for FPGA synthesis.

4.4 Code Generation for Hardware Implementation

4.4.1 From Objects to Hardware Blocks

Whilst the implementation of the MEP semantics in a software system is quite straightforward, as demonstrated in the next section, the inherently different properties of hardware systems must be considered when applying the MEP approach here. The following paragraphs discuss those properties and outline the semantics we employ to derive a hardware implementation directly from an MEP based specification.

Instantiation. In the context of hardware generation we first have to clarify the actual semantics of class instantiation. Our approach maps class instances to hardware blocks whilst their attributes map to registers. Operations are synthesized to equivalent logic for each individual instance. It is thus not possible to instantiate a hardware class at run time[2] and all instances must be determined at compile time. Thus we define all instantiations in the constructors. Consequently all Instances except the system's root Instance are owned through composition and form a composition tree at compile time. A shared Association between Instances defines non-owned links between instances, e.g., for delegation. However, those Instances have a common transitive owner in the composition tree.

For our hardware generation all constructors are evaluated starting from the constructor of the system main class. This results in the creation of the corresponding hardware blocks and their interconnections based on the parameterized constructor calls for owned instances and interconnection definitions in the constructors.

The evaluation of the constructors can be done simply by running a software simulation that computes the initial configuration of the hardware system. To eliminate the need for a complete simulation, Operation calls other than constructor calls within the constructors can be prohibited. However, this is not a significant limitation because Operation calls may be used later to initialize the instance at run time.

[2]Not considering run time reconfiguration for specific FPGAs here.

Note here that destructors are essentially normal Operations in the context of hardware synthesis where the implied deallocation is meaningless. Anyway, we prescribe that no access to an instance is allowed after the destructor call.

Size and Granularity Considerations. The mapping from similarly sized specifications to hardware may yield circuits which differ significantly in size. A very simple example is an adder compared to a multiplier when both are specified with the same size. Furthermore, unlike in software systems, hardware instances contain individual copies of each circuit. It is therefore essential for hardware synthesis to reflect these properties in the structural part of the specification to make efficient use of the available silicon. This usually leads to the creation of reusable elements at a finer granularity than in software systems.

Finally, the bit width of data is an important factor for the size and depth of a circuit. Unlike in software systems, which use fixed size registers, hardware systems directly implement logic for each bit of the operand size. Thus it is essential to support this at the design level through explicit operand size definition.

Connecting Hardware Blocks. Dynamic links between software Instances are just pointers, which are essentially without cost in software systems. In hardware, dynamic links are costly and require significantly more effort to be implemented, e.g., by the use of multiplexers.

In our approach hard wired links between hardware blocks can be defined in three ways. The first way is through composition and they are defined through nested constructor calls. The second alternative is to pass references to connected instances to the constructor, which are assigned to final attributes. Finally, the explicit declaration of hard wired interconnections at an owning Instance is possible. We support this through parameterized Activities defining fixed data and control flows between a subset of the owned instances. The example in Fig. 4.11 takes the third alternative and defines the interconnection of the Butterfly Instances through the parameterized IDCTActivity. A parameterized Activity is a template defining the fixed interconnection between elements from the given Instances resulting in a named Activity. The newly defined Activity can be referenced from other Activities (see Fig. 4.12).

We investigated the support of dynamic hardware links to enable polymorphism during execution as an essential feature. From a hardware point of view dynamic links connect one hardware block to other hardware blocks with a common interface. This enables an Instance to access all Instances of a given type und its subtypes including all Attributes, Operations, and their parameters. However, since dynamic hardware links can be highly resource consuming, e.g., by their implementation through cascaded multiplexers, their individual application has to be carefully considered in the context of hardware synthesis.

Multiplexing. Multiplexing of hardware units is used for multiple access to single resources. We model this by associating a class implementing the multiplexed unit. However, access to an instance has to be controlled in order to avoid reentrancy problems.

Reentrancy. When moving directly from software to hardware design, the non-reentrancy of hardware needs special attention. Operations in software systems are usually reentrant since software systems typically use the stack to store the execution state during an Operation call. In hardware systems this state is the state of the respective circuit. Thus recursion or concurrent invocation of the same Operation leads to erroneous behavior. Therefore we require that all hardware Operations must be declared as non-reentrant.

Interconnecting IPs. When composing components, e.g., IPs, to a system, we have to consider the corresponding instances und their interconnected interfaces. This makes it possible to define the complete system through a UML Component Diagram where the individual components are Instances of Classes defined by their exposed and required Interfaces. Note that this is semantically equivalent to a subset of the Class Diagrams we use at the top level.

4.4.2 Handel-C Code Generation

Handel-C is a procedural C based Hardware Description Language (HDL) for FPGA synthesis. In Handel-C, variables represent registers and internal and external memory. Control constructs with C based expressions, i.e., conditional branches and loops, define control flows. For synthesis the Celoxica tools transform the nested Handel-C procedures into corresponding gates, flip-flops and additional components, like multipliers.

For hardware code generation we define a transformation from MEP based specifications into Handel-C programs. This is mainly a straightforward translation where control oriented behavior models like Activities are directly mapped to Handel-C control constructs. The translation of StateMachines involves a transformation into an explicit implementation, which is well understood in hardware synthesis. However, Handel-C, like other commonly used synthesizable HDLs, is not object-oriented. Therefore we need to implement the object-oriented MEP semantics through procedural programs.

For that our code generation takes two passes. During the first pass all relevant information is collected from the design model and stored in a temporary data structure, which we call the instance model. Based on the instance model linear generation of Handel-C is performed (see Fig. 4.7). The first pass starts from the class representing the complete system. That class inherently has just one instance, which we call the *system instance*. The *instance model* with the system instance as a root represents the actual system in terms of Instances and

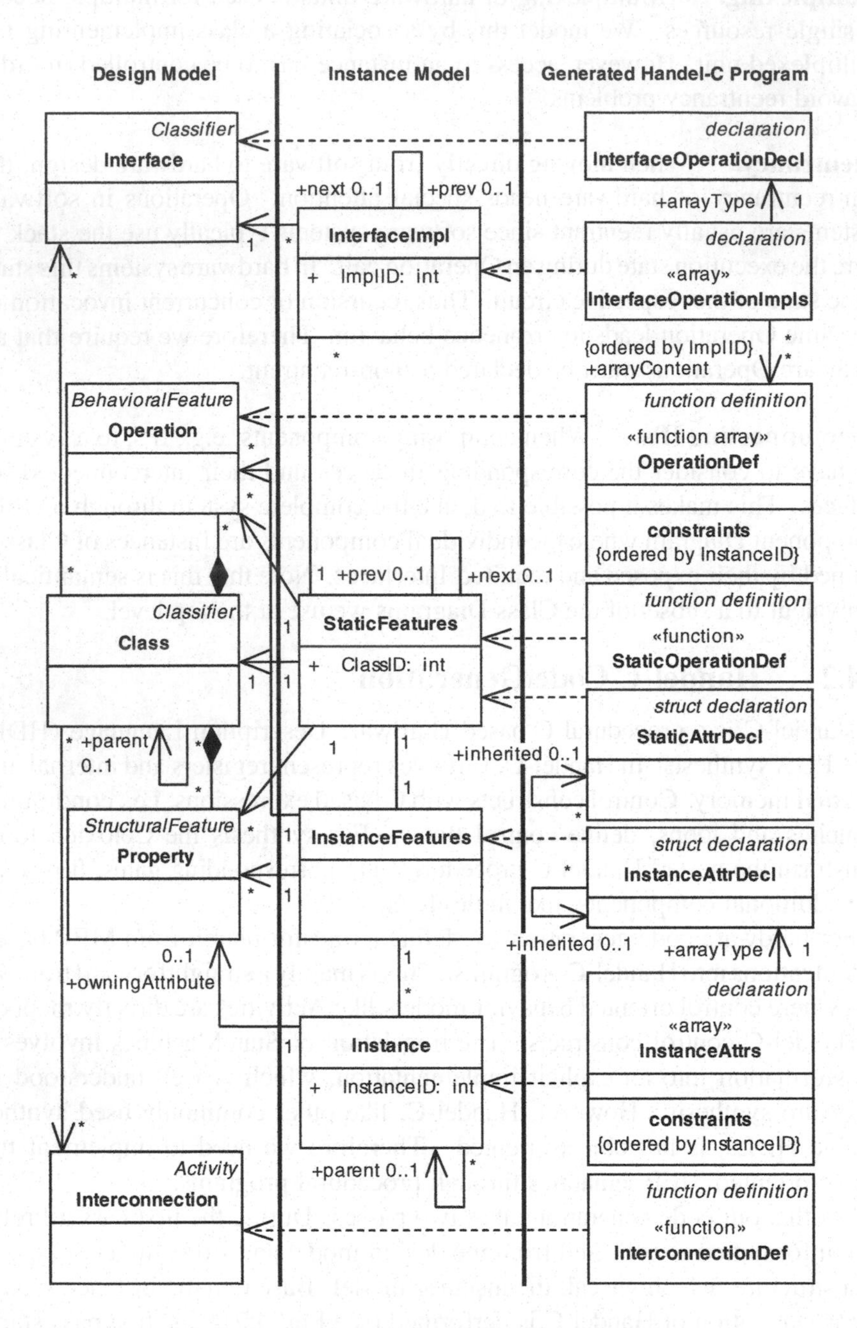

Figure 4.7. Handel-C Code Generation Overview

their Classifiers. It provides navigation capabilities to directly access the individual model elements during code generation. The instance model is derived from the design model by complete traversal along the constructor call hierarchy starting from the constructor of the system instance. The respective elements from the design model are included through directed links. The list of all Classes is available through the StaticFeatures element in the instance model. For each Interface a list of all implementations is accessible through InterfaceImpl. Finally, all Instances can be derived through the link to InstanceFeatures. The lists are also used to assign consecutive identifiers to the respective elements. The identifiers are required during code generation as well as during run time to identify the actual type of an Instance to implement polymorphism.

The generated code essentially consists of Handel-C records and functions. Attributes are represented as nested records as defined by the class hierarchy. Furthermore, static and non-static attributes are separated into two different record hierarchies. Final Attributes are defined directly as constants using fully qualified names.

The behavioral parts of the MEP specification are resembled by Handel-C functions. An individual function is required for each Activity instance contained in each Operation for each Instance as well as for each static Operation.

For Operation calls with fixed targets, the corresponding function is invoked directly. However, this cannot be applied for Parameters or dynamically associated Instances of a Class or Interface. In those cases the actual Instance and the respective Handel-C function implementing the Operation may vary. Therefore the call is implemented through a function table. As pointed out earlier, this solution can be very resource consuming.

Recall that the MEP defines an Operation by a StateMachine. That StateMachine is directly implemented by a Handel-C function as a event loop, which processes the StateMachine events. The loop implements each state as a case of a switch statement which handles the individual events and executes the do, entry, exit, and effect Activities by calling the corresponding Handel-C functions. When an Operation is defined by a trivial StateMachine with a single Activity, the function for the Activitiy is directly implemented without generating the redundant code for the StateMachine.

Each Activity instance is implemented as a separate function. This also applies to the parameterized Activities for interconnecting Instances in the constructor. Because we target for hardware we can fully exploit the concurrency given by the forks and joins of Activities. They are directly mapped to parallel and sequential Handel-C blocks. The code generation for Activities is based on a list scheduling algorithm where code for the model elements of an Activity is generated in the order in which these elements are enabled through the data and control flows of the Activity.

4.4.3 IDCT Hardware Example

As outlined in the previous section, the Inverse Discrete Cosine Transformation (IDCT) is one of the steps most critical to performance in the MPEG decoding process, so that we take it for our hardware code generation example. During this step, a complete video frame is decoded. It is composed from blocks of 8x8 bytes representing either chrominance or luminance information for a block of 8x8 pixels. The same decoding is applied to each of the blocks.

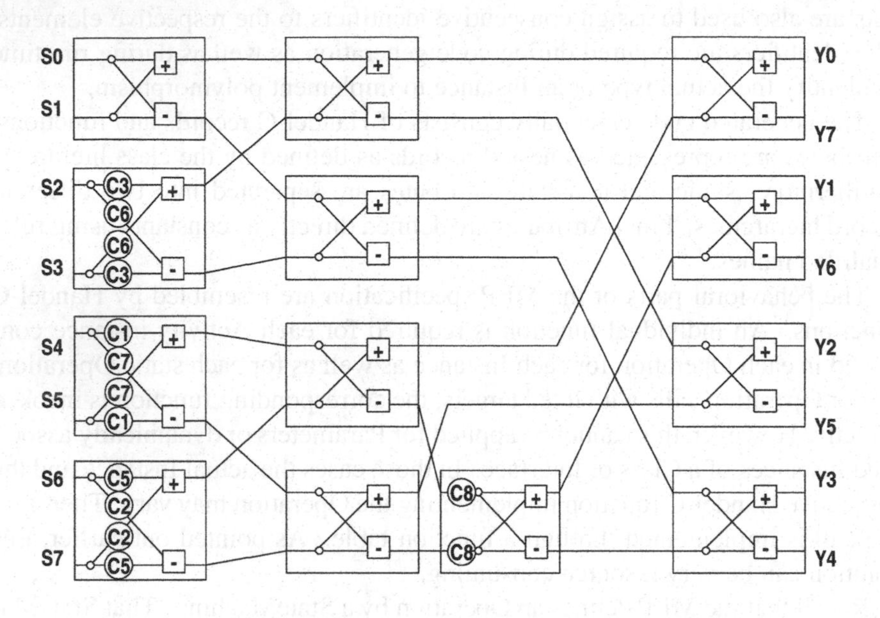

Figure 4.8. IDCT Implementation using the Chen-Wang Algorithm

We apply the Chen-Wang algorithm [227, 35] to decode an 8x8 byte block of a frame through the application of a butterfly network (see Fig. 4.8) to each row and column of an 8x8 byte block. For that, the Chen-Wang algorithm employs 13 butterfly components for processing a row of 8 pixels with 8 bits each. Each butterfly component has two inputs and two outputs. The inputs are connected to an addition and a subtraction to produce the outputs. This also involves a constant multiplication in some instances. We take the Chen-Wang algorithm to demonstrate an efficient hardware implementation of the IDCT by means of Handel-C code generation in the context of our MEP approach. That complements the MPEG decoder example, which was introduced in Subsection 4.3.3.

For hardware implementation we introduce a HardwareIDCT class implementing the IDCT interface as defined in the previous Class Diagram of the MPEG decoder (see Fig. 4.4). The MEP specification for the HardwareIDCT component starts with a Class Diagram describing the Features and composition of that Class (see Fig. 4.9).

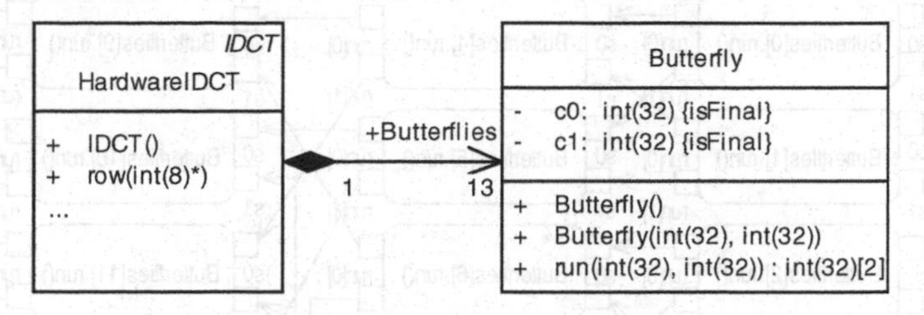

Figure 4.9. HardwareIDCT Class Diagram

As an implementation of the Chen-Wang algorithm the HardwareIDCT class is composed from 13 instances of a Butterfly Class using scaled integers with 32-bit precision. The Butterfly Class essentially contains constructors for initializing the multiplier constants and a run() Operation for computing the output of the particular butterfly component for a given set of inputs.

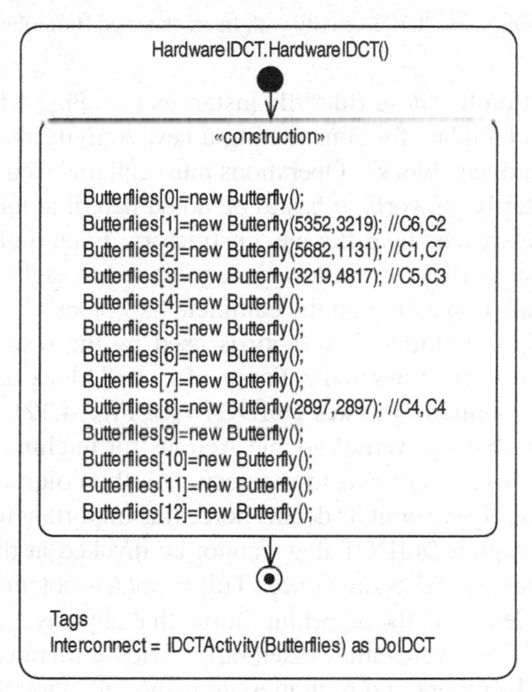

Figure 4.10. Constructor Creating and Interconnecting Butterflies

The HardwareIDCT class (see Fig. 4.10) defines all owned Butterfly instances through respective constructor calls in its constructor. This constructor also applies a parameterized Activity identified as IDCTActivity to define a

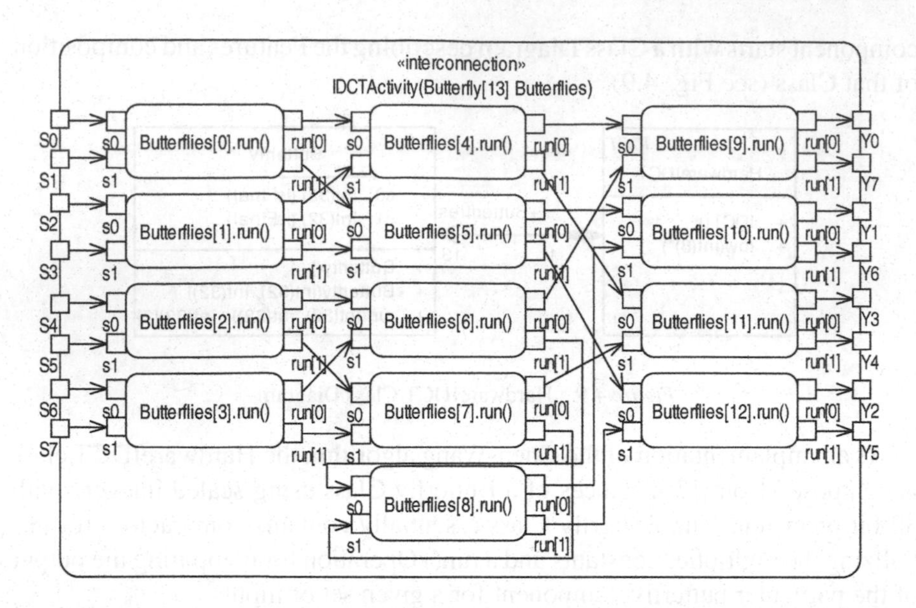

Figure 4.11. IDCTActivity with Interconnected Butterflies

fixed interconnection for those Butterfly instances (see Fig. 4.11). This Activity is applied as a template for constructing a new Activity that represents the interconnected hardware blocks. Operations must call this Activity as DoIDCT to invoke the butterfly network. It has to be noted here that the parameterized Activity can be referenced by different constructors. Even multiple references within the same constructor are allowed, e.g., to create eight copies of IDCTActivity for parallel operation on the complete 8x8 block.

In our example the DoIDCT Activity is used by the row() and column() Operations to process the rows and columns of an 8x8 block respectively. The row() Operation essentially invokes DoIDCT (see Fig. 4.12). Additionally it contains ReadVariable and WriteVariable Actions for fetching and storing the input and output bytes from and to memory. As the column() Operation is implemented similarly we omit its details here. It is important to note that since both operations employ DoIDCT they cannot be invoked at the same time as the respective hardware exists only once. This is not a problem in our example, as the data dependences in the algorithm imply that anyway.

The Handel-C Code generation essentially yields a hierarchy of functions representing the Operations and Activities along the call hierarchy starting from the system main function. To illustrate code generation for our example, we focus on the previously introduced parts of the specification.

Recall that the HardwareIDCT constructor instantiates 13 Butterfly Instances (see Fig. 4.10), which results in interconnections defined by IDCTActivity (see Fig. 4.11). The corresponding Handel-C function `DoIDCT0()`, as given in Fig.

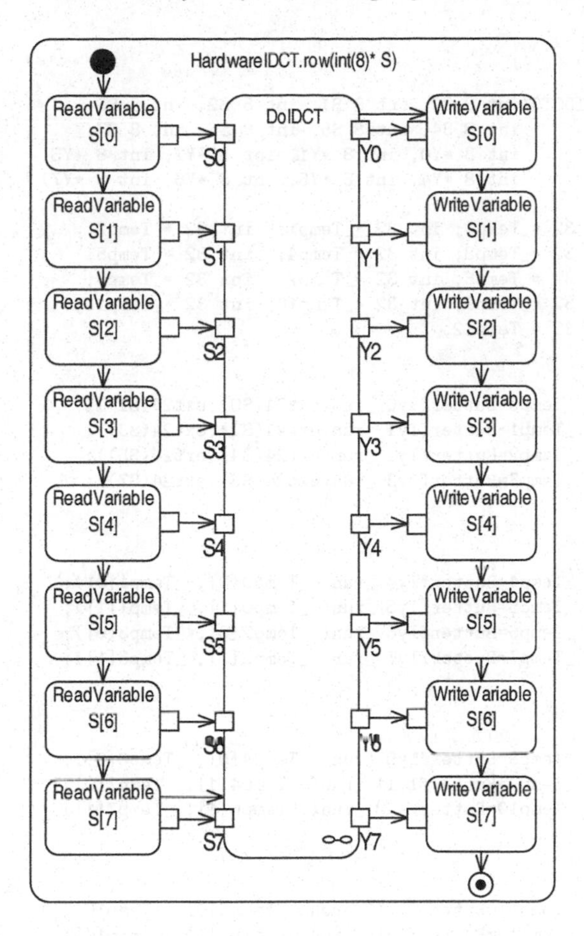

Figure 4.12. AD for `HardwareIDCT.row()` with Reference to `DoIDCT`

4.13, calls the different instances of the Butterfly and implements the data flow through temporary variables. The signature of the function is determined by the Pins of IDCTActivity where OutputPins are mapped to parameters passed by reference. The actual sequence of statements is determined by the list scheduling algorithm of the code generation. Parallel parts of the IDCTActivity are simply mapped to parallel blocks.

The `HardwareIDCT0_row()` function implementing the row() Operation, as given in Fig. 4.14, is generated similarly. That function essentially invokes the `DoIDCT0()` function implementing IDCTActivity for the same instance after sequentially fetching the operands. The result is sequentially stored back to memory as defined by the Activity.

Code generation for each Butterfly instance is straightforward. As an example, Fig. 4.15 gives the implemention of `Butterfly.run()` for the instance

```
void DoIDCT0(int 8 S0, int 8 S1, int 8 S2, int 8 S3,
             int 8 S4, int 8 S5, int 8 S6, int 8 S7,
             int 8 *Y0, int 8 *Y1, int 8 *Y2, int 8 *Y3,
             int 8 *Y4, int 8 *Y5, int 8 *Y6, int 8 *Y7)
{
    int 32 *_Temp0; int 32 *_Temp1;  int 32 *_Temp2;
    int 32 *_Temp3; int 32 *_Temp4;  int 32 *_Temp5;
    int 32 *_Temp6; int 32 *_Temp7;  int 32 *_Temp8;
    int 32 *_Temp9; int 32 *_Temp10; int 32 *_Temp11;
    int 32 *_Temp12;
    par
    {
        _Temp0=Butterfly0 _run(ext24(S0),ext24(S1));
        _Temp1=Butterfly1 _run(ext24(S2),ext24(S3));
        _Temp2=Butterfly2 _run(ext24(S4),ext24(S5));
        _Temp3=Butterfly3 _run(ext24(S6),ext24(S7));
    }
    par
    {
        _Temp4=Butterfly4 _run( _Temp0[0], _Temp1[0]);
        _Temp5=Butterfly5 _run( _Temp0[1], _Temp1[1]);
        _Temp6=Butterfly6 _run( _Temp2[0], _Temp3[0]);
        _Temp7=Butterfly7 _run( _Temp2[1], _Temp3[1]);
    }
    par
    {
        _Temp8=Butterfly9 _run( _Temp4[0], _Temp6[0]);
        _Temp9=Butterfly11 _run( _Temp4[1], _Temp7[0]);
        _Temp10=Butterfly8 _run( _Temp6[1], _Temp7[1]);
    }
    par
    {
        _Temp11=Butterfly10 _run( _Temp5[0], _Temp10[0]);
        _Temp12=Butterfly12 _run( _Temp5[1], _Temp10[1]);
        *Y0= _Temp8[0]<-8;
        *Y3= _Temp9[0]<-8;
        *Y4= _Temp9[1]<-8;
        *Y7= _Temp8[1]<-8;
    }
    par
    {
        *Y1= _Temp11[0]<-8;
        *Y2= _Temp12[0]<-8;
        *Y5= _Temp12[1]<-8;
        *Y6= _Temp11[1]<-8;
    }
}
```

Figure 4.13. Handel-C Code for `DoIDCT0()`

```
void HardwareIDCT0 _row(int 8 *S)
{
    int 8 _Temp0;  int 8  _Temp1; int 8  _Temp2;
    int 8 _Temp3;  int 8  _Temp4; int 8  _Temp5;
    int 8 _Temp6;  int 8  _Temp7; int 8  _Temp8;
    int 8 _Temp9;  int 8  _Temp10; int 8  _Temp11;
    int 8 _Temp12; int 8  _Temp13; int 8  _Temp14;
    int 8 _Temp15;
    _Temp0=S[0];
    _Temp1=S[1];
    _Temp2=S[2];
    _Temp3=S[3];
    _Temp4=S[4];
    _Temp5=S[5];
    _Temp6=S[6];
    _Temp7=S[7];
    DoIDCT0(_Temp0, _Temp1, _Temp2, _Temp3,
            _Temp4, _Temp5, _Temp6, _Temp7,
            _Temp8, _Temp9, _Temp10, _Temp11,
            _Temp12, _Temp13, _Temp14, _Temp15);
    S[0]= _Temp8;
    S[1]= _Temp9;
    S[2]= _Temp10;
    S[3]= _Temp11;
    S[4]= _Temp12;
    S[5]= _Temp13;
    S[6]= _Temp14;
    S[7]= _Temp15;
}
```

Figure 4.14. Handel-C Code for `HardwareIDCT0_row()`

with index 0. The generated code is basically given by a parallel Handel-C block. The actual calculations there involve the final instance variables c0 and c1 of the Butterfly Instance. They are initialized in the Butterfly constructor. This also underlines that Handel-C function arrays are not sufficient to represent multiple instances of a function for different Class instances, as the individual instance variables are directly accessed.

```
int 32 *Butterfly0 _run(int 32 s0, int 32 s1)
{
   int 32  _Result[2];
   par
   {
      _Result[0]=Butterfly0 _c0*s0+Butterfly0 _c1*s1;
      _Result[1]=Butterfly0 _c1*s0-Butterfly0 _c0*s1;
   }
   return  _Result;
}
```

Figure 4.15. Handel-C Code for `Butterfly_run()` with Index 0

4.5 The MEP Virtual Machine

The previous section introduced concepts to generate synthesizable hardware descriptions from the MEP based UML models. This section presents an alternative approach implementing the MEP semantics by a Virtual Machine (VM) for embedded real time systems. Here we just sketch the basic concepts and refer the reader to [185] for further details.

The MEP VM introduces an alternative run time environment for embedded software which supports interrupts, and timeouts. First evaluations of a VM prototype implemented on an FPGA have shown that it is possible to reach a performance level in the region of C programs [185].

The virtual machine executes binary encoded specifications which resemble the object-oriented structure of the MEP-based UML specification. It enables efficient execution of StateMachines with Activities, which are compiled to a microprocessor-like byte code with object-oriented extensions (see Fig. 4.16). In contrast to the Java Virtual Machine [122], the MEP VM is based on the execution of state oriented models. For efficient execution in the context of reactive systems, the MEP VM byte code has specific instructions for state transitions and advanced event processing, like timeouts and interrupts. Additionally, the architecture separates the scheduler and the memory manager from the byte code and ESM interpreter, so that it can be easily tailored to the individual application and platform.

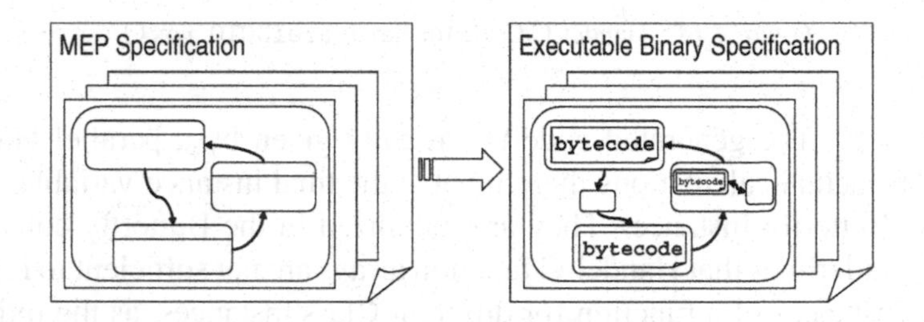

Figure 4.16. Transformation to Binary Specifications

The *Model Execution Unit* (MEU) and the *Runtime Kernel* (RK) are the core elements of the MEP VM as shown in Fig. 4.17. The MEU directly executes MEP specifications. The RK is implemented as an executable specification. It includes memory management and thread scheduling. Additionally it incorporates bootstrapping, e.g., for loading the initial executable UML model.

The Model Execution Unit is composed of two interacting interpreters for the different parts of the binary UML representation; one for the Executable StateMachines (ESM) and another one for the byte code. A separate timer manages hard timeouts at the precision of a millisecond.

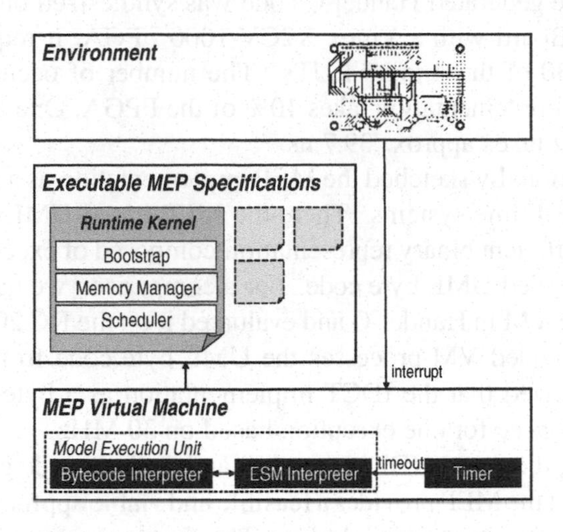

Figure 4.17. MEP VM Architecture

The ESM Interpreter performs immediate state transitions when events occur. Six predefined events are supported: Completion, Interrupt, Timeout, Division by Zero, Overflow, and Out of Memory. When a state is entered, or a transition is executed, the interpreter first checks if an Activity with byte code is defined. If there is byte code in the embedded Activity, the byte code interpreter is invoked. A generated byte code sequence typically finishes with a special COMPLETE instruction, which generates a completion event. If no Activity is defined or when the final state is reached, an immediate completion event is generated.

The byte code Interpreter executes the byte code in a microprocessor like manner using instruction scheduling. To provide a hardware-independent execution platform, the interpreter has no registers. All variables are managed on the stack. The overall conception of the byte code Interpreter is comparable to the Java VM, but is embedded in the execution of ESMs and relies on the customizable scheduling and memory management provided by the Runtime Kernel.

4.6 Conclusions

We have introduced our Model Execution Platform (MEP), which is based on the notion of executable UML and bridges the gap from specification to implementation. The MEP introduces concepts for true object-oriented, model-based system design. We start from Class, StateMachine, and Activity Diagrams, arriving at software or a synthesizable Hardware Description Language such as Handel-C. We have demonstrated the Handel-C code generation from Class and

Activity Diagrams by the example of an Inverse Discrete Cosinus Transformation (IDCT). The generated Handel-C code was synthesized on a Celoxica RC 200 evaluation Board with a Virtex X2CV-1000 FPGA. It required 461 slice flip-flops and 730 of the 4 input LUTs. The number of occupied slices was 550. The IDCT implementation takes 10% of the FPGA. One execution of the IDCT at 30 MHz takes approx. 39.7 ns.

We have additionally sketched the MEP implementation as a virtual machine for embedded real time systems. There the MEP-based UML specification is compiled to an efficient binary representation composed of executable state machine with embedded UML byte code. For a comparison we have additionally implemented the VM in Handel-C and evaluated it on the RC 200 board, where the hardware encoded VM processes the UML byte code from the SDRAM. Final figures showed that the IDCT implementation as a byte code takes approximately 103 μsec for one execution based on 30 MHz.

Both variants, the direct HDL generation and the byte code generation, have demonstrated that the MEP provides a feasible and viable approach for hardware and embedded software systems design. The first evaluations give promising results for both variants. However, evaluations of more complex examples and projects are still necessary in order to draw more general conclusions.

Chapter 5

Hardware/Software Codesign of Reconfigurable Architectures Using UML

Bernd Steinbach,[1] Dominik Fröhlich,[1,2] Thomas Beierlein[2]

[1] *Institute of Computer Science*
Technische Universität Bergakademie Freiberg
Freiberg, Germany

[2] *Institute of Automation Technology*
Hochschule Mittweida (FH) - University of Applied Sciences
Mittweida, Germany

Abstract The development of systems comprising hardware and software components has been a demanding and complex problem. To manage the informational and architectural complexity inherent to these systems novel approaches are taken. In this chapter we present an approach that is based on the concepts of model driven architecture, platform based design, and hardware/software codesign.

5.1 Introduction

Reconfigurable architectures are a relatively novel means of constructing computer systems. These architectures comprise one or more microprocessors and reconfigurable logic resources. The microprocessors execute the global control flow and those parts of the application that are uncritical to the overall performance. The logic resources act as coprocessors and execute the performance-critical algorithms of the system or specialized input/output operations. In *runtime reconfigurable architectures* (RTR) the logic resources are reconfigurable while the system is in operation. RTR systems feature the dynamic adaption of the functionality executed by the coprocessors to the current requirements of the application.

The main applications of reconfigurable architectures are *system on chip* (SoC) and high performance computing. Reconfigurable systems enable the acceleration of the overall system whilst cutting development and manufactur-

G. Martin and W. Müller (eds.), UML for SOC Design, 89–117.

ing costs. In contrast to classical ASIC approaches the functionality implemented in hardware is not fixed, it may change even while the system is already deployed. This makes them an enabling technology for SoC prototyping and evolving systems.

The development of applications of reconfigurable architectures is a very complex and error-prone task. The most current development approaches, which are based on programming languages or mixed languages, are insufficient owing to their strong focus on implementation and the inherent technology dependence [14]. Thus novel directions must be taken. In this chapter we present a development approach that is based on the *Unified Modeling Language* (UML) and a dedicated action language. UML 2.0 and the action language are used for the object-oriented system specification and design at the *system level* of abstraction.

The approach is backed by a dedicated tool called the MOCCA compiler (Model Compiler for reconfigurable architectures). Given complete and precise models of the applications design and the design and implementation platforms, MOCCA can automatically perform partitioning, estimation, and the implementation of the system into hardware/software modules. The synthesized implementation is directly executable and exploits the capabilities of the hardware architecture. The key concepts of object-orientation — inheritance, polymorphism, encapsulation, and information hiding — are preserved down to the implementation level. Thus the archetypal break in paradigms, languages, and tools of current object oriented hardware development efforts is avoided.

The full automation of the implementation process supports early verification and simulation. The time required for the implementation of a system level specification is cut down by orders of magnitude. As a result, the focus is shifted from implementation towards modeling. This offers tremendous gains in productivity and quality. The synthesized hardware and software modules fit together by definition because the compiler has automatically implemented them. The change of the system design, the algorithms, the partitioning, and the implementation platform is encouraged.

The rest of this chapter is structured as follows. In Section 5.2 we introduce our development approach. We describe the used models, artifacts, and the utilization of UML and the action language. In Section 5.3 we define how the UML system design models are mapped into implementations. The focus of this chapter is on the mapping of implementation models into hardware/software implementations, which is described in detail in Section 5.4. The approach is illustrated by AudioPaK an encoder/decoder for the lossless compression of audio streams [84]. In Section 5.5 experimental results for the example are presented and then this chapter is concluded.

5.2 A Platform Model Driven Development Approach

5.2.1 Overview

In this section a brief overview of the general development approach will be given. Because of the lack of space we focus on the key concepts and artifacts. The methodology is not described here, a thorough discussion can be found in [14][203].

The development of applications is based on platforms, whereas different platforms are used for design, implementation, and deployment. As a result, a strict separation of development concerns is accomplished. Moreover, this approach eases validation, portability, adaptability, and reuse. Platforms have been used ever since in the areas of software and hardware development. However, platforms are mostly captured implicitly in language reference manuals, libraries, and tools, which hampers their automated interpretation by computers.

Platforms represent sets of assumptions, which are the foundation of any development effort. In our approach, these assumptions are made explicit by platform models, whereas each platform is specified by a dedicated platform model. Platform models abstract from the details of the platform described, but carry enough information to avoid iterations in the design flow. They are the basis for the definition of application models that describe a certain aspect of the system under development. The relationship between the platform models and application models is illustrated in Figure 5.1. The platform models define the space of applications which may be developed with the respective platforms. Each particular set of application specific models represent one point in the application space. Different platform models normally share application models.

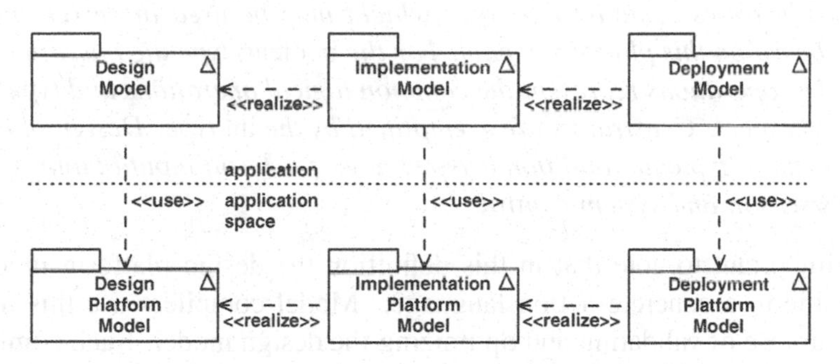

Figure 5.1. Relationships between Models

All models are described using UML 2.0 [154] and a dedicated action language called MAL (MOCCA Action Language). MAL was developed to enable the detailed specification of behavior in UML models in computation intensive and control intensive applications. This language is compliant to the UML action semantic specification. It has a medium level of abstraction because it requires the developer to make data organization and data access explicit. However, this allows us to employ standard analysis, optimization, and estimation techniques [14].

In the following sections we discuss the employed platforms and models and give some brief examples of their content and meaning.

5.2.2 Design Model and Design Platform Model

A *design model* defines an executable and implementation independent realization of the use cases of a system. This model defines the structure and behavior of the system. System structure is defined by UML packages, classes and interfaces and their various relationships. For concurrency specification active classes are supported. System behavior is defined by operations and state machines. Detailed behavior is defined by UML actions, whereas MAL is used as action language. Concurrent control flows are synchronized by guarded operations.

Each design model is based on a design platform which is specified by a *design platform model.* The content of a design platform model depends on the targeted application domain. It specifies the types, constraints, and relationships used for system design. For each type the relationship to other types in terms of generalizations and dependencies, the supported operations, and constraints are defined. The definition of the basic types is mandatory when developing systems using UML because the types defined by the UML specification are defined too loosely in order to be useful in real world designs.

EXAMPLE 5.1 *Figure 5.2 shows a small part of a design platform model. The example illustrates some design types whicht may be used in design models that are based on this platform model. For the* boolean *type the operations are shown. The operations represent the common logical operations and type casts one would expect. Constraints are exemplified by the* int *type. Design platform models contain typically additional types, e.g., for basic input/output, access to actors/sensor, and system control.*

It is important to note that in this definition the design platform model is not specific to a concrete action language. Model compilers use this model for the purpose of validating and optimizing the design model. Such compilers make only minimum assumptions about types. The validity of a design model is determined entirely by the design platform model and the UML well formedness rules. Designers may add new types and operations to the design platform model which are treated by model compilers as primitive types. For these elements the designer may then provide optimized implementations in the implementation platform.

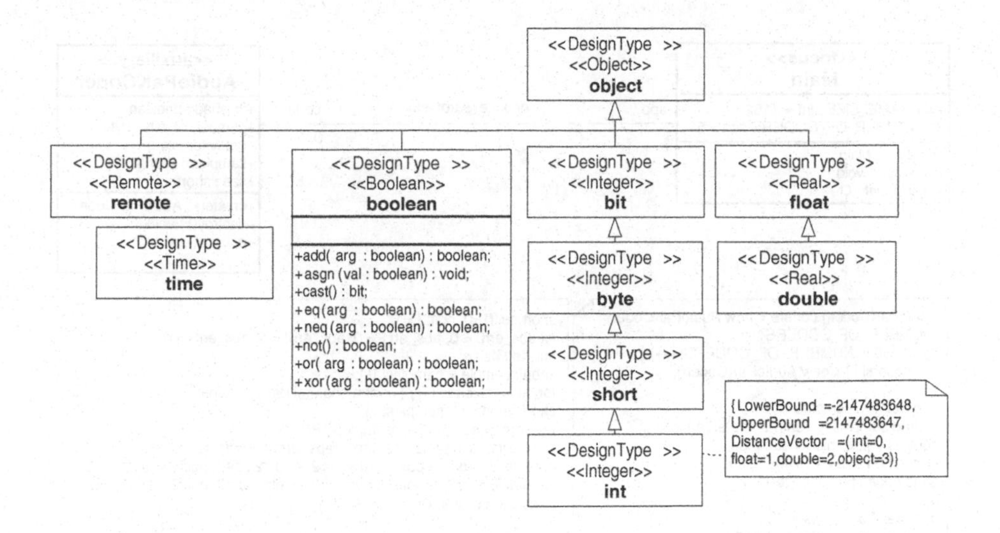

Figure 5.2. Design Platform Model Example

EXAMPLE 5.2 *Figure 5.3 shows the design model of an AudioPaKCoder [84]. AudioPaK is a well known algorithm for lossless compression of audio information. In the example the class* `Main` *instantiates a number of AudioPaKCoder objects, which encode frames of audio samples, whereas each sample is represented by a 16 Bit integer. The frames are written to coder objects which concurrently encode them while* `main` *performs other tasks, such as input/output, filtering, etc.. The example shows the intra channel decorrelation algorithm performed by the operation* `encode`.

5.2.3 Implementation Model and Implementation Platform Model

An *implementation model* defines the realization of a design model in terms of implementation classes, components, artifacts and relationships. This model has the same functionality as the design model; however, it typically realizes this functionality differently. Implementation models define how structure and behavior are realized with the services provided by the implementation platform. There are many implementation models for a given design model. An implementation model is derived from a design model by applying a sequence of transformations and mappings (see Section 5.3.4). The implementation model is created manually, or (semi-) automatically by model compilers.

Each implementation model is based on a specific implementation platform. Implementation platforms define the realization of design platforms, whereas

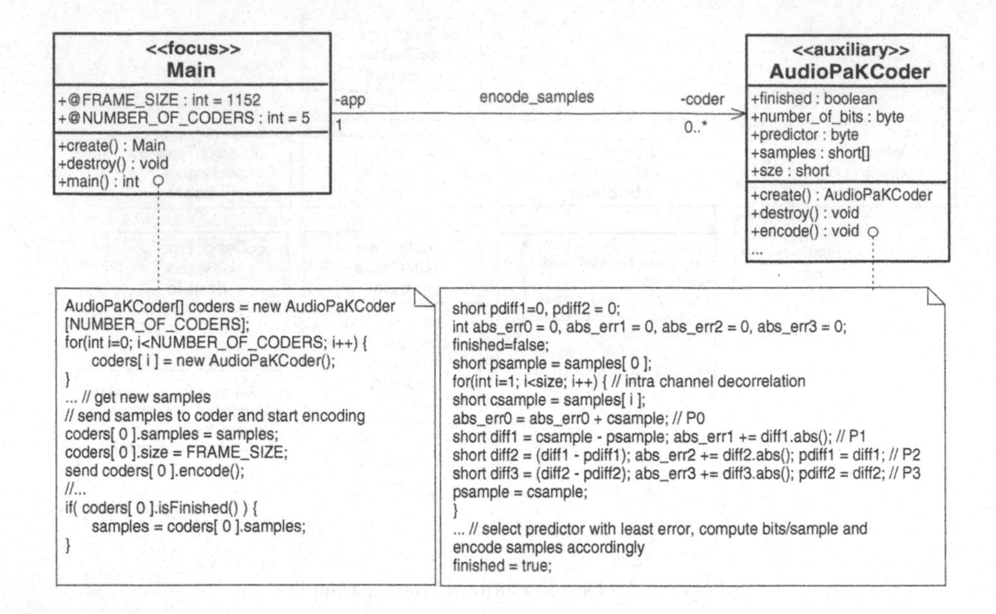

<<focus>>
Main
+@FRAME_SIZE : int = 1152
+@NUMBER_OF_CODERS : int = 5
+create() : Main
+destroy() : void
+main() : int
-app
1
encode_samples
-coder
0..*
<<auxiliary>>
AudioPaKCoder
+finished : boolean
+number_of_bits : byte
+predictor : byte
+samples : short[]
+sze : short
+create() : AudioPaKCoder
+destroy() : void
+encode() : void
...
AudioPaKCoder[] coders = new AudioPaKCoder
[NUMBER_OF_CODERS];
for(int i=0; i<NUMBER_OF_CODERS; i++) {
coders[i] = new AudioPaKCoder();
}
... // get new samples
// send samples to coder and start encoding
coders[0].samples = samples;
coders[0].size = FRAME_SIZE;
send coders[0].encode();
//...
if(coders[0].isFinished()) {
samples = coders[0].samples;
}
short pdiff1=0, pdiff2 = 0;
int abs_err0 = 0, abs_err1 = 0, abs_err2 = 0, abs_err3 = 0;
finished=false;
short psample = samples[0];
for(int i=1; i<size; i++) { // intra channel decorrelation
short csample = samples[i];
abs_err0 = abs_err0 + csample; // P0
short diff1 = csample - psample; abs_err1 += diff1.abs(); // P1
short diff2 = (diff1 - pdiff1); abs_err2 += diff2.abs(); pdiff1 = diff1; // P2
short diff3 = (diff2 - pdiff2); abs_err3 += diff3.abs(); pdiff2 = diff2; // P3
psample = csample;
}
... // select predictor with least error, compute bits/sample and
encode samples accordingly
finished = true;

Figure 5.3. Design Model Example: AudioPaK Coder

each implementation platform realizes one design platform. Each implementation platform is specified by an *implementation platform model.* For each processing element in the hardware platform an implementation platform model is defined.

An implementation platform model is the set of types, constraints, transformations, and tools that may be used for the realization of design models. As design platforms, implementation platforms are defined on the basis of abstract instruction sets[1] and their execution characteristics. This model is used by model compilers to perform high level design space exploration (platform mapping), estimation, and synthesis.

EXAMPLE 5.3 *Figure 5.4 shows a part of an implementation platform model for a HDL implementation platform. The diagram depicts design types from Example 5.1 and the implementation types used for their realization. The implementation types and operations are characterized by quality of service and generation information.*

[1] An instruction may be the native instruction of microprocessors, the operations directly implementable in hardware, but also high level operations of programming languages.

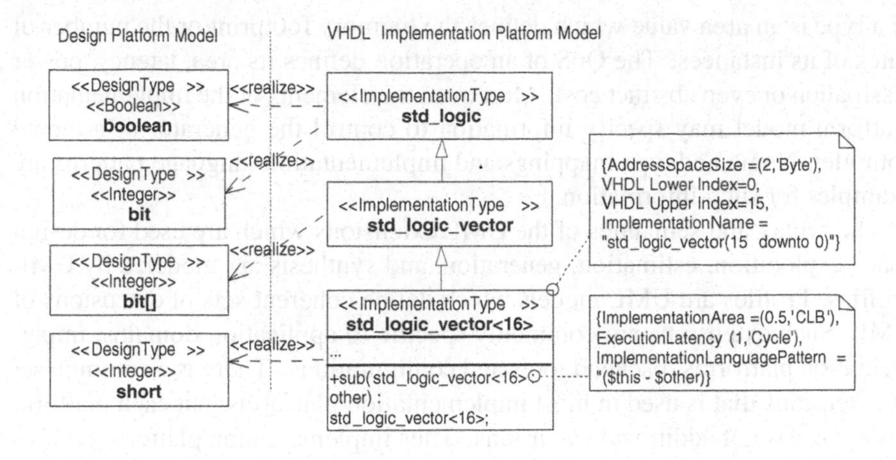

Figure 5.4. Implementation Platform Model: Types and Mappings

EXAMPLE 5.4 *Figure 5.5 continues the presentation of the VHDL implementation platform model. Implementation platforms may contain pre-implemented hardware/software modules. Model compilers can integrate such cores in the generated modules. The example shows a clock generator component, a PCI bridge, and a storage component that may be used in hardware designs. The components specify a number of interfaces for clocking, local/external access and interconnections, which are specified in detail by interfaces and classes. The components are implemented in VHDL, the UML model just specifies the necessary information to integrate them into hardware modules.*

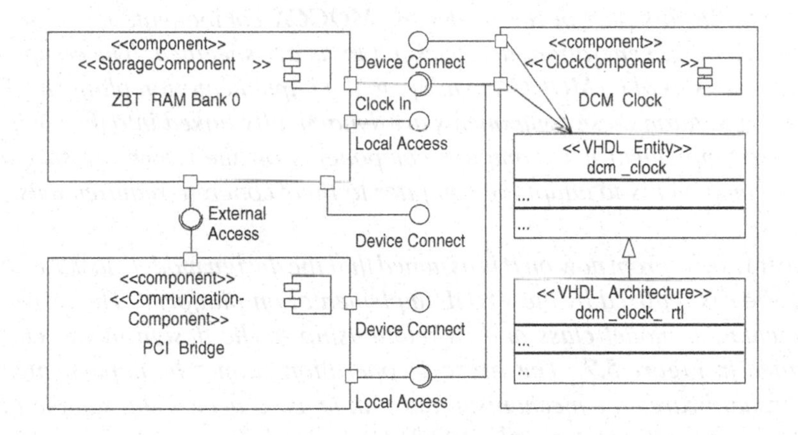

Figure 5.5. Implementation Platform Model: IP Integration

The implementation components, types, and operations are characterized by their quality of service (QoS). As shown in Figure 5.4, the QoS characteristic of a type is an area value which defines the memory footprint or the number of gates of its instances. The QoS of an operation defines its area, latency, power dissipation or even abstract cost. Moreover, the elements of the implementation platform model may specify information to control the generator of a model compiler. Name and type mappings and implementation language patterns are examples for such information.

The syntax and semantics of the UML extensions which are used for design space exploration, estimation, generation, and synthesis are modeled by UML profiles. Profiles are UML models which define coherent sets of extensions of UML. Such extensions are commonly specific to application domains, implementation platforms, backend tools and configurations. There is a common set of extensions that is used in most implementation platforms, but each platform has its own set of additional extensions. Thus implementation platform profiles are defined hierarchically.

UML extensions can be interpreted by users and model compilers. In order to avoid design iterations, implementation models and implementation platform models must reflect the characteristics of the compiled/synthesized hardware/software artifacts as close as possible. Thus it is important to give the model compiler control over the implementation process. Owing to the huge variety of implementation platforms a model compiler for SoC should be able to adapt to the set of platforms being used. To make this adaption convenient and straightforward, the respective components of the model compiler are modeled in the implementation platform model.

EXAMPLE 5.5 *The approach to model the modeling compiler components in the implementation platform model is exemplified for the MOCCA compiler in Figure 5.6. In this part of the model the MOCCA components used for estimation, mapping, generation, and backend tools are specified. The component specification is used by MOCCA to adapt to the implementation platform. During the compilation these components are dynamically linked into the compiler. Users may implement new compiler components on their own, or specialize existing components to adapt the compiler to their concrete requirements.*

EXAMPLE 5.6 *From now on it is assumed that the design model class* `AudioPaKCoder` *is realized by the VHDL implementation platform. The respective implementation model class and its relationship to the design model class is illustrated in Figure 5.7. The* `encode` *operation cannot be implemented directly in hardware; the mechanism and protocols to access the sample array must be made explicit. A simple WISHBONE-like bus interface is used in the example [197]. The behavior of* `encode` *is transformed respectively to access the samples through this interface. This model will be refined in the course of this chapter.*

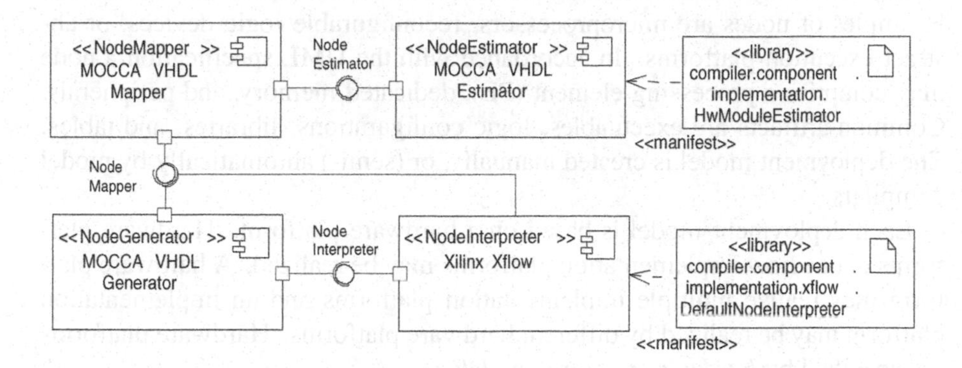

Figure 5.6. Implementation Platform Model: Compiler Components

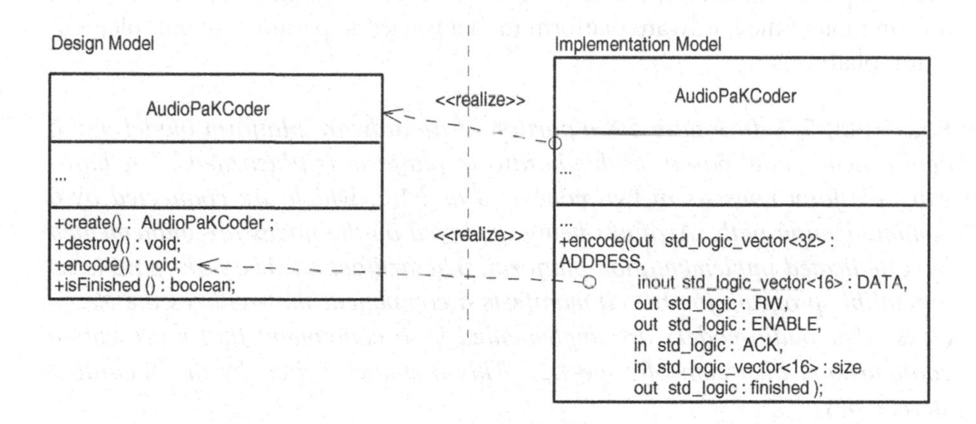

Figure 5.7. Implementation Model: AudioPaKCoder Mapping

5.2.4 Deployment Model and Hardware Platform Model

A *deployment model* defines the deployment of an implementation model of the application on a target hardware architecture. This model defines the deployment relationship between the nodes of a hardware architecture and the artifacts which manifest the components of the implementation model. As a result, the deployment model fixes the execution of the implementation model. Examples of nodes are microprocessors, reconfigurable logic devices, or abstract execution platforms. In accordance with the UML specification a node may comprise a processing element (PE), dedicated memory, and peripherals. Common artifacts are executables, logic configurations, libraries, and tables. The deployment model is created manually, or (semi-) automatically by model compilers.

Each deployment model is based on a hardware platform. Hardware platforms define how implementation platforms may be realized. A hardware platform may realize multiple implementation platforms and an implementation platform may be realized by different hardware platforms. Hardware platforms are specified by *hardware platform models*.

A hardware platform model defines the nodes, communication paths, and constraints of a hardware architecture. Hardware platforms commonly do not specify the micro-architecture of hardware nodes; they define the services provided by the hardware resources. For instance, the number of logic and memory resources, clock rate ranges, scheduling policies, communication protocols are specified by constraints. The hardware platform model must contain just enough information to enable high quality design space exploration. The constraint representation is similar to the QoS in the implementation platform. This information of the hardware platform model is used to parameterize implementation platforms.

EXAMPLE 5.7 *In Figure 5.8 a portion of a hardware platform model and a deployment model based on this hardware platform is illustrated. The hardware platform consists of two nodes* `h0` *and* `h1`*, which are connected by a communication path. Artifacts being deployed on the nodes are implemented by a dedicated implementation platform. The artifact* `audiopak.exe` *is an executable program for* `h0`*. It manifests a component that realizes the* `Main` *class. The audio coders are implemented by a component that represents a configuration bitstream of node* `h1`*. This is made explicit by the according stereotypes.*

The deployment and implementation related models complement each other. The deployment platform model and implementation platform model are referred to as *target platform model*. The implementation model and deployment model are subsumed in the *platform specific model* [14]. A design model is

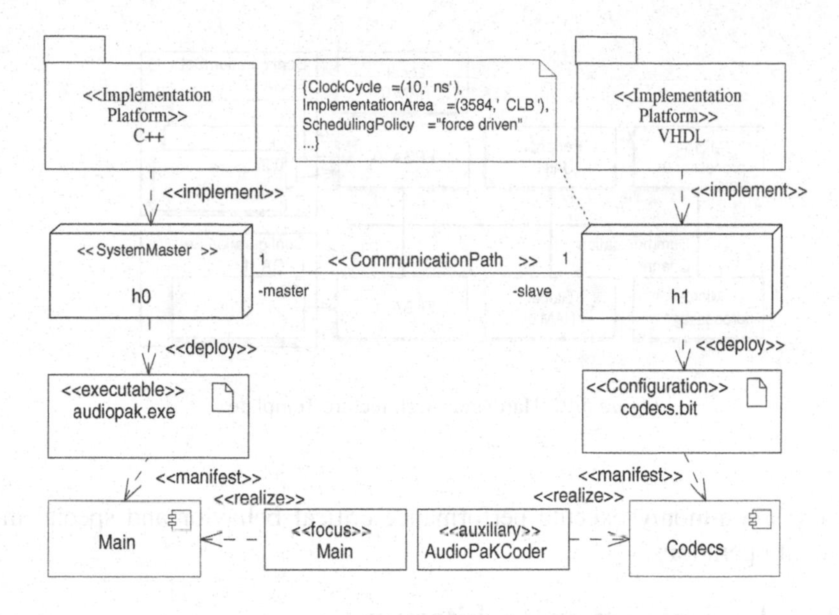

Figure 5.8. Deployment Model

implemented on a new target platform by using an according target platform model. The design model and design platform model are not required to change.

5.3 Mapping Design Models onto Implementation Models

5.3.1 Hardware Architecture

The mapping of a design model to an implementation model depends on the physical and logical system architecture. The physical system architecture is determined by the architecture of the underlying hardware. The development approach targets SoCs with heterogeneous multiprocessor architectures that are complemented by reconfigurable logic resources. Figure 5.9 shows an architectural template for these architectures.

The hardware is a heterogeneous multiprocessor system that comprises a number of processing elements, realized as microprocessors or FPGAs (field programmable gate arrays). Each FPGA is associated with a set of configurations. Runtime reconfigurable FPGAs are associated with multiple configurations which are activated on demand. The nodes of the hardware architecture are connected through a common communication channel. The PEs may possess local communication channels to reduce contention on the global channel.

A dedicated PE acts as the system master. This node is commonly a microprocessor. It controls the overall control flow of the system, invokes functionality implemented by the slaves, and triggers the reconfiguration of the FPGAs.

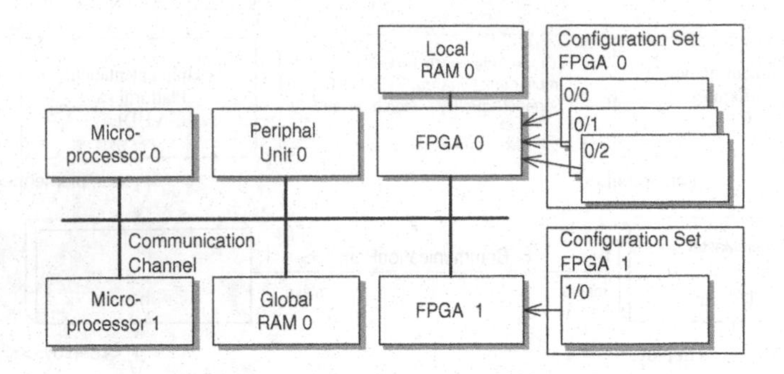

Figure 5.9. Hardware Architecture Template

The slaves commonly execute performance-critical behavior and special input/output operations.

5.3.2 Logical System Architecture

The logical system architecture is implemented with the resources of the hardware architecture. This architecture orients towards the object model of computation [57]. The system functionality is realized by objects communicating through structured messages. Each object has a state (the data encapsulated by the object) and offers services which may be accessed through the interface of the object. The services define the object behavior. A service is invoked by sending an appropriate message to the object. The respective service handler may change the state of the object and it may also send messages to other objects. Messages may be sent synchronously or asynchronously.

Objects are realized with PEs and memory resources. Objects of the same class may be executed on multiple PEs and may have different PE specific implementations. The messages are transmitted through the common communication channel or a local communication channel.

5.3.3 Design Space Exploration

During design space exploration a partition of the system function amongst the components of the target hardware architecture is computed. The implementation space, as defined by the target platform model, is explored for feasible alternatives implementing the system. The quality of each alternative being explored is estimated. The partition which optimizes the performance and satisfies the design constraints best is chosen for implementation. As a result, an implementation model and deployment model for the design model is computed.

Design space exploration can be performed fully automated by model compilers or based on models which have been partially partitioned by the designer. Of course, designers may also create the implementation and deployment model manually. The algorithms for design space exploration and hardware/software partitioning are chosen depending on the application domain and the model compiler. Model compilers for hardware/software systems are extendable to support user-specific partitioning algorithms. This degree of freedom in choosing the partitioning mode and algorithms is possible owing to the application of a common language. This is one of the strengths of UML based codesign of hardware/software systems.

Common object-oriented UML specifications are not directly implementable on the given target, due to the polymorphism and use of dynamic data structures. Thus during design space exploration those parts of the design model that are not implementable on a target platform must be transformed respectively. Design models are transformed to enable implementation and to optimize implementation and execution characteristics. The transformations are manifested in the implementation and deployment models.

5.3.4 Model Transformations

Model transformations are used to map and optimize design models. Mapping transformations are used during design space exploration to define the realization of design models. If a design model is not directly implementable model compilers search for a sufficient set of transformations. If no such transformation set exists the design model is not implementable with the implementation platform. There are three classes of transformations: allocations, behavior transformations, and structure transformations [203]. Allocation operators act on all elements of design and implementation models. These operators map model elements to the target platform by assigning sufficient sets of resource services.

EXAMPLE 5.8 *Figure 5.10 continues the implementation model of Example 5.6. For the action* `diff1=csample-psample` *in the operation* `encode` *the allocation of resource services is demonstrated. The respective activity group is shown in a compact notation, according to [146]. For the realization of the "-" operation, a sufficient resource service is allocated. The service is implemented by a* `sub` *operation of the VHDL type* `std_logic_vector<16>`*. Other examples for this transformation type are allocations of storage and communication components as shown in Figure 5.5.*

Behavior transformations act on UML behavior specifications, e.g., state machines, activities and actions, and the model elements which implement them. Structure transformations act on model elements from which the structure of a UML model is built.

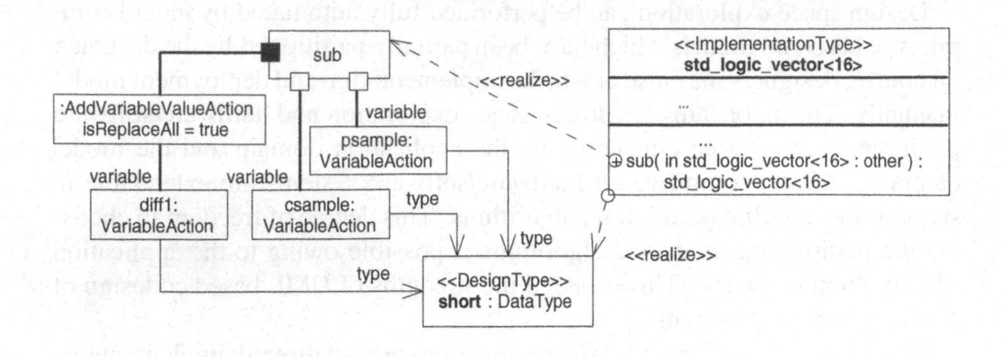

Figure 5.10. Allocation Transformation Example

EXAMPLE 5.9 *An example for behavioral transformations has already been given in Example 5.6. The interface and behavior of the* encode *operation are transformed in such a way that the samples are accessed via a bus interface. Similar transformations are necessary to accomplish message transfers in hardware modules. Another example of this kind of transformation are (de) compositions and optimizations, such as arithmetic/logic optimizations, dead code elimination, and constant value propagation. Behavior transformations are often accompanied by respective structure transformations.*

In the same way as programming language compilers and synthesis tools, model compilers may use transformations to optimize the implementation and execution characteristics of design and implementation models. In principle the most basic and advanced optimization techniques that are applicable to common object-oriented specifications are applicable to UML models as well, and there is good reason to do so. In contrast to the final implementation executable UML models describe the entire control and data flow of an application in a single and consistent representation. This information is used by advanced model compilers to perform more aggressive optimizations than would be possible on the final implementation. Examples are dead code elimination, constant and value range propagation optimizations that act globally on UML models [138]. Also optimizations known from behavioral synthesis are applicable.

5.4 Mapping Implementation Models onto Platform-Specific Implementations

5.4.1 Overview

The implementation of the system relates directly to the structure and behavior of the implementation model. The object-oriented properties of the design model are preserved in the final implementation. Classes, interfaces, operations, and attributes of implemented objects map directly to their counterparts in the implementation model. Inheritance and polymorphism are preserved independently of the realization! In addition, the components and artifacts as specified in the implementation model are preserved in the implementation. Artifacts stereotyped as 'Configuration' map to configuration bitstreams of (reconfigurable) logic resources. The instantiation of a component manifested by such an artifact corresponds to the loading of the bitstream into the physical device. The same relationship applies for 'executable' artifacts and microprocessor devices.

For both hardware and software realizations, the question of the right level of abstraction of the final implementation and a suitable language arises. In an object-oriented approach it seems quite natural to choose a language supporting the object paradigm for software implementation. This approach makes the implementation convenient and straightforward. The final compilation is delegated to 3rd party compilers. However, as a result a fair amount of control over the final implementation is lost. In performance and resource critical applications, this uncertainty can cause iterations in the development flow. To avoid this problem model compilers for critical application domains may generate microprocessor specific assembly language implementations. For the purpose of this chapter C++ is used to implement software modules.

For the implementation of hardware modules the same considerations as for software apply. Owing to the tight timing and resource constraints imposed by the hardware it is even more important to reflect the implementation model directly in the implementation. In principle the implementation can be delegated to behavioral synthesis tools [58]. However, as for software implementations it is hardly possible to compute good estimates of the synthesized results. Moreover, the programming language based approaches, such as SPARK and Forge [134][231], are restricted by the employed languages and the directly synthesizable language subsets of the targeted HDLs. Thus model compilers for SoC synthesize hardware modules directly from UML models on the register transfer level (RTL). For the purpose of this chapter, the implementation of hardware modules is described with synchronous VHDL-RTL designs.

The direct implementation of components, classes, and features is convenient and straightforward. Whereas in software this is a well understood problem, in hardware implementations this approach raises the following challenges:

Dynamic Object Instantiation/Destruction Owing to the static nature of even partially reconfigurable hardware, the efficient instantiation and destruction of hardware objects is not possible efficiently. The class instantiation per reconfigurable device is by far too expensive in terms of the number of required logic resources and reconfiguration time.

Polymorphic Features Polymorphism is an important property of object-oriented specifications. It should be supported directly by hardware implementations. Current approaches avoid polymorphism by prohibiting inheritance or overriding of behaviors.

Communication of Objects Objects should be able to communicate independent from their realization. In mixed software and hardware implementations no single, common mechanism for message exchange exists.

In the following sections the mapping of implementation models to hardware/software implementations is discussed. Owing to the focus of this book the main focus is on hardware implementations. Solutions of the stated problem areas are presented.

5.4.2 Software Implementation

The implementation model of a design determines the implementation of the design. Model compilers implement hardware/software modules that realize the same function and structure as the implementation model. Owing to different implementation patterns and styles multiple implementations are possible for an implementation model. These differences are reflected in the QoS in the implementation platform model so that it does not affect the quality of the design space exploration results.

The implementation patterns and rules are either manifested in the respective components of the model compiler or in code generation annotations in the UML meta model. The latter approach is taken by xtUML [131]. It has the advantage of being defined entirely with UML models and dedicated generation languages (archetypal language). However, it orients towards single language software implementations. Design space exploration, estimation, (automated) model transformations, and mixed language implementations are not directly supported. Thus in our approach the former approach is taken.

The classes of the implementation model being deployed on microprocessor nodes are directly implemented in C++. Local proxy objects manage the communication between local and remote objects. For each remote object that is accessed by a local object a proxy is instantiated locally. The proxy encapsulates the communication mechanism. Thus the objects of an application are not required to share a common address space. The proxy is explicitly modelled in the implementation platform model as an *remote* type (see Figure 5.2). As a result, the model compiler can compute high quality estimates of the characteristics of distributed applications.

Objects which are implemented in reconfigurable hardware are managed by a dedicated service called *RTR Manager* [174]. This manager encapsulates the specifics of the reconfigurable hardware, e.g., reconfiguration modes, input and output functions, and communication. The most important task of this service is to process application requests for the creation and destruction of hardware objects. Hardware objects are created and destroyed on demand. An application that instantiates an hardware object requests it from the RTR Manager by its type. The RTR Manager searches for a suitable object in the currently instantiated bitstreams and serves its proxy to the application. If no bitstream containing an object of the searched type is currently instantiated, the RTR Manager dynamically instantiates an appropriate bitstream.

EXAMPLE 5.10 *Figure 5.11 illustrates the basic architecture of the AudioPaK Example. The instance* `main` *of class* `Main` *and a number of proxies for hardware objects are implemented in software. The actual instances of the class* `AudioPaKCoder` *are realized by means of reconfigurable resources. Each hardware object is accessed from software through a dedicated proxy. The proxies are served to the application by an instance of the* `RTR Manager` *service.*

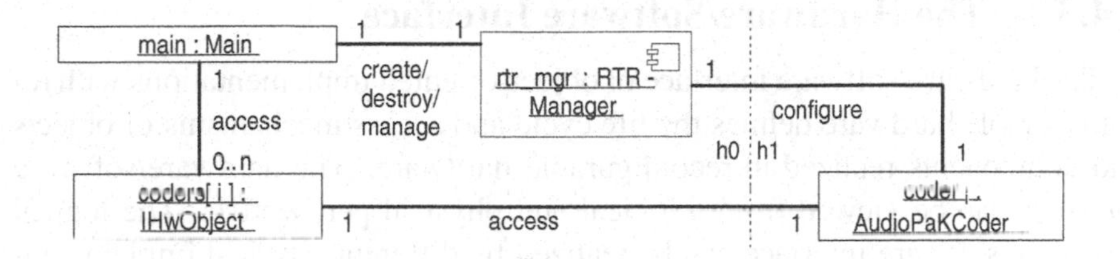

Figure 5.11. Software Architecture of AudioPaK Example

Proxies can be used directly in the software implementation to provide a simple yet fast mechanism for accessing the hardware objects. Alternatively, proxies may be wrapped by software implementations of the hardware object classes. If an instance of the software object is created the object tries to instantiate its hardware counterpart. In case of success the hardware object is used, otherwise the software object switches transparently to the software implementation. This approach also enables the transparent migration between hardware and software objects. Advanced model compilers generate such implementations automatically. Implementations are guaranteed to be correct because the compiler has generated them.

EXAMPLE 5.11 *Figure 5.12 shows a part of the operation* `Main::main`. *In the loop, a number of hardware objects will be created. Recall from Example 5.2 that the algorithm writes one frame with audio samples to the first coder object and starts it asynchronously. In the course of the algorithm it is checked whether the encoding is finished, and, if so, the encoded frame is read.*

```
...
coders = new smartptr<IHwObject>[ 5 ];
for(i = 0; (i < 5); i = i + 1) { // create coder objects
    coders[ i ] = RTRManager::getInstance()->createObject( 0 );
}
... // get new samples
coders[ 0 ]->write<short>(8, samples, 1152);
coders[ 0 ]->write<short>(2312, 1152);
coders[ 0 ]->start<char>(4, 2);  // start encode (async)
... // fill other coders
coders[ 0 ]->execute<char>(4, 1); // isFinished (sync)
if( coders[ 0 ]->read<bool>(2315) ) {
    samples = ((short*)((int) coders[ 0 ].getPtr() + 8));
} ...
```

Figure 5.12. Software Implementation of `Main::main`

5.4.3 The Hardware/Software Interface

The hardware/software interface of object-oriented implementations with reconfigurable hardware defines the life cycle and access mechanisms of objects and components realized in reconfigurable hardware. The hardware/software interface can be viewed from a logical and physical perspective. The logical hardware/software interface can be realized by different physical implementations. The concrete implementation depends upon the target platform and the model compiler.

For efficiency reasons the life cycle of hardware objects is different from the life cycle of software objects. In order to avoid costly reconfigurations hardware objects are reused as much as possible. Because a true dynamic instantiation/destruction of objects is not efficiently possible in hardware, these objects are pre-instantiated at compilation time and synthesized into configuration bitstreams. The objects are dynamically allocated on demand; the RTR Manager serves as object broker. Additionally, in RTR systems the objects and hardware configurations are dynamically bound to logic resources.

This mechanism is reflected in the life cycle of objects and bitstreams, which is illustrated in Figure 5.13. Because of the tight relationship between the hardware objects and their configuration context, as the container of the hardware objects, both life cycles influence each other. Each object and its bitstream will go through three states, `X_UNBOUND`, `X_BOUND` and `X_ALLOCATED` (where `X` is either `OBJ` for hardware objects or `BS` for bitstreams). As long as the bitstream is not loaded into the reconfigurable hardware, the bitstream and the contained objects are in state `X_UNBOUND`. When a bitstream is loaded it changes its state to `BS_BOUND`. The objects contained go to the state `OBJ_BOUND`. Objects allocated by the application change go from state `OBJ_BOUND` to `OBJ_ALLOCATED`.

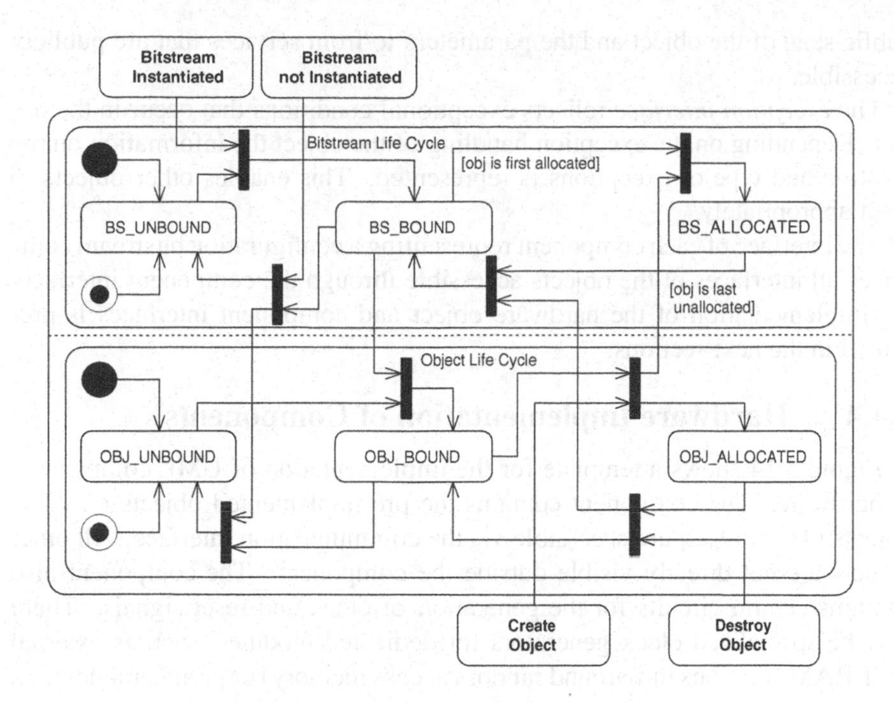

Figure 5.13. Object and Configuration Bitstream Life Cycles

All objects returned by the application will set their state back to `OBJ_BOUND`. The last object of a context returned causes the bitstream to be set back to `BS_BOUND`. Until the bitstream is unloaded from the hardware, the objects will still be available for allocation. The bitstream is not allowed to be unloaded from the hardware as long as it is in the state `BS_ALLOCATED`.

The mechanisms for the access of objects are defined by the object interfaces. The interface of each object consists of a control interface, a data interface, and an exception interface.

The *control interface* allows for identification, typing, and access to the object. Objects are uniquely identified in their object space. The ID represents the address of the object and is set during initialization. This field is only required if the object address must be made explicit in the object interface. The type field represents the dynamic type of the object. It is used to select the proper implementations of polymorphic features. This field is only required if the object may have different dynamic types. The message field uniquely identifies the type of service accessed by a message sent to the object. The service which is executed in response to the message may depend on the dynamic object type. The message parameters are passed through the data interface.

The *data interface* allows one to access the object state and to pass input/output parameters from and to objects. The interface contains the entire public state of the object and the parameters to/from services that are publicly accessible.

The *exception interface* reflects exceptional conditions that occur in the object. Depending on the exception handling of the object the information on the position and type of exceptions is represented. This enables other objects to react appropriately.

The interface of each component representing a configuration bitstream comprises all interfaces of the objects accessible through the component interface. An implementation of the hardware object and component interfaces is presented in the next sections.

5.4.4 Hardware Implementation of Components

Figure 5.14 shows a template for the implementation of UML components in hardware. The component contains the pre-implemented objects O_i. The objects $O_0, ..., O_{n+4}$ are accessible via the communication interface. All other objects are not directly visible outside the component. The component also contains central circuits for the generation of clock and reset signals. There may be specialized clock generators for dedicated modules, such as external ZBT RAM (zero bus turnaround random access memory) or peripheral devices.

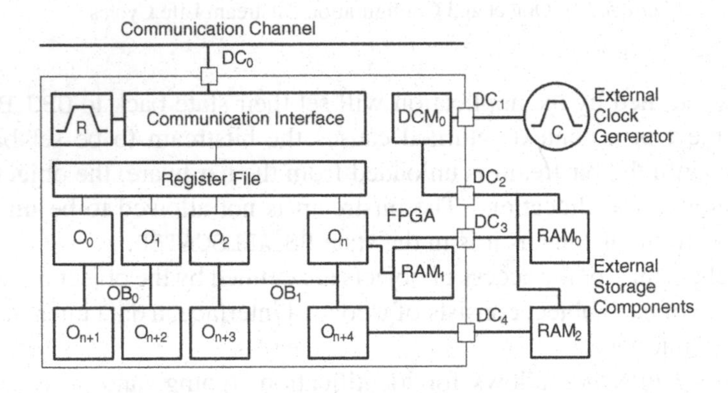

Figure 5.14. Hardware Component Template

The ports DC_i are manifestations of the `Device Connect` interfaces as modeled in the implementation platform model (see Figure 5.5). This interface type is used to establish connections between internal logic resources and external devices through the physical device interface. From the UML specification of these interfaces the model compiler generates such implementations. The generation process is controlled by the generation information in the model. In the VHDL implementation the interface of the top level design module comprises of all signals of the DC ports.

Additionally required components modeled in the implementation platform model are instantiated on demand. For instance, if RAM_0 was realized with `ZBT RAM Bank 0` (Figure 5.5), a `DCM Clock` (Digital Clock Manager) component is instantiated (DCM_0) and connected to RAM_0. The interfaces of both components are modeled by UML interfaces and classes. The logic using the storage is connected to its `Local Access` interface. The adaption of the external to the local interface is realized by an appropriate VHDL wrapper component. The reset and clock signals of the design are automatically generated by the model compiler. If the number of sink flip-flops on the clock tree exceeds the maximum fan-out of the clock generator, a clock generator component is instantiated to buffer the signal.

Multiple objects are clustered in a hardware configuration. The number and type of the hardware objects being clustered in a single configuration is determined either manually or automatically. For this the global message exchange of classes and their object creation/destruction characteristics is analyzed. The number of concurrently required object instances is estimated from real or estimated execution profiles of the application [52].

The public interface of publicly visible objects is realized by a register file. The register file allows one to access the control, data, and exception interfaces of the respective hardware objects. Collaborating objects communicate via direct connections or object buses (OB_i). In order to minimize bus contention there may be multiple object buses in one component. During design space exploration the model compiler tries to identify reasonable groups of collaborating objects. For each object group an object bus is generated which connects all member objects of the group.

Only the publicly visible objects can be instantiated by the software portion of the application. The other objects are hidden from the outside and are used as helper objects within the component. The access to an object of a given type must be independent of the object template and the dynamic object type. In order to accomplish this the public object interfaces of all objects of a given type and all of its subtypes must be identical. The realization of polymorphism is hidden behind the external object interface.

Model compilers for SoC perform the interface layout during generation. In the real layout alignment constraints on the items in the register file, which are imposed by the communication channels, are considered. Model compilers assign to each element in the register file an address that satisfiess the alignment constraints in the system. In addition the corresponding address decoders are automatically generated. Software modules accessing a hardware object use only the relative addresses of the member elements of the object interfaces. The software proxies are parameterized with the absolute object address by the RTR Manager. The proxies compute the absolute address of an element when it is accessed. This ensures that software and hardware always fit together and that the object access is independent of real object addresses and implementations.

EXAMPLE 5.12 *Figure 5.15 illustrates the implementation of a hardware component for our design example. In this implementation a PCI bus (Periphal Component Interconnect) is used as communication channel. The PCI bridge from our implementation platform (see Figure 5.5) is used to adapt the external bus to the internal bus* `Local Access`*. The external interface of the component was modeled as* `Device Connect` *UML interface. A register file is connected to the internal bus. The address decoders and address range decoders AD/ARD_i select the appropriate register from the file at each PCI access. The* `AudioPakCoder` *objects are also connected to the register file. Their control interface is implemented in the address decoders and the control register. The other registers realize the data interface; the BRAMs (Block RAMs) have been modeled in the implementation platform, the other registers are generated by the model compiler. Owing to the flat class hierarchy of the example no type register is required. The exception interface is empty because no exceptions are thrown in the example.*

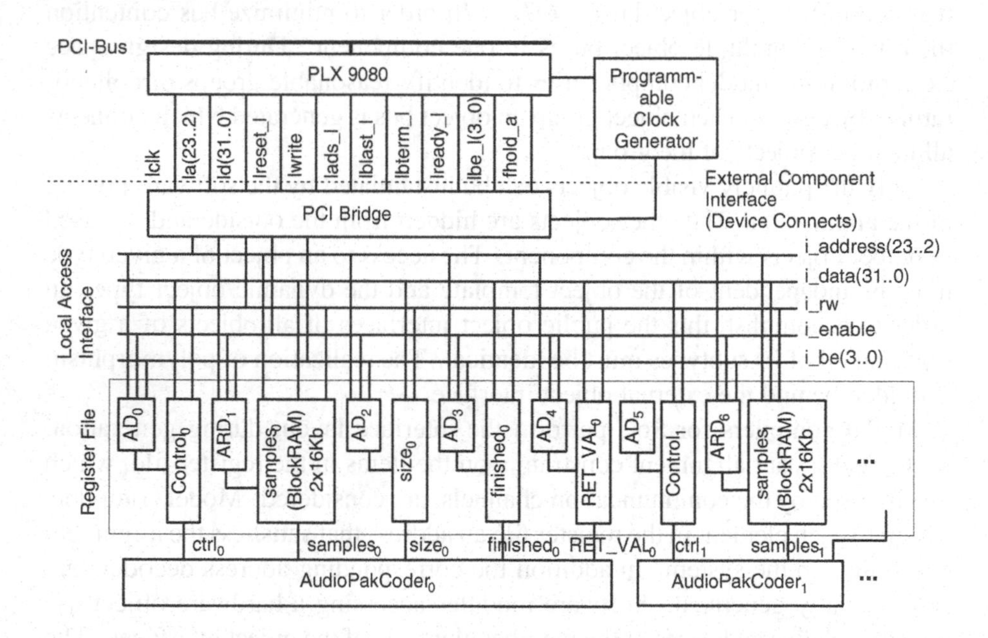

Figure 5.15. Component Implementation AudioPaK Example

5.4.5 Hardware Implementation of Objects

Figure 5.16 shows a template for the implementation of objects in hardware. The template shows the implementation of one object. The object contains the pre-implemented operation O_i, the operation parameters P_i and the attributes A_i. The object can be of type T_0 or T_1. The sets P_i and A_i comprise the data interface of the object. Operation O_2 has a non-empty exception interface.

For each type the visible features must be provided in the interface. The set of visible operations is a superset of the operations that are defined by a type. Model compilers automatically eliminate unused features before design space exploration.

A control and type register implements the control interface. The type register holds the current dynamic type of the object. The control register contains the two signals *GO* and *DONE* for each operation. An operation is started by setting its *GO* signal. The end of execution is signaled by the operation when it sets its *DONE* signal.

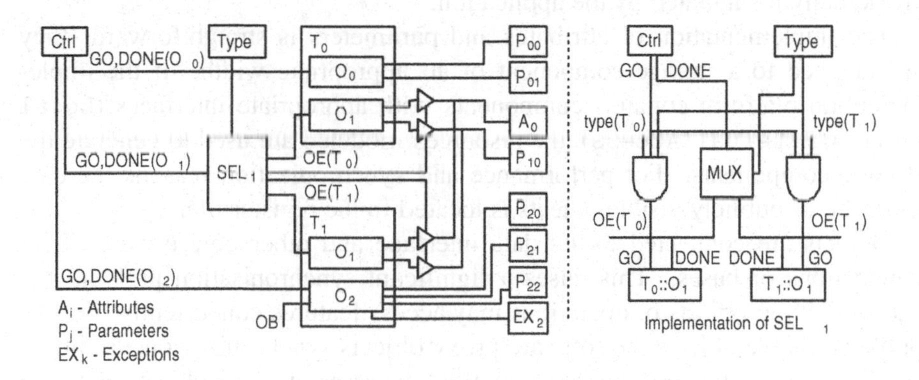

Figure 5.16. Hardware Object Template

Operation O_1 is polymorphic, that is, the behavior to be executed when the operation is started depends on the current object type. The selection of the behavior to execute is performed by a selector circuit whose implementation is depicted on the right hand side of Figure 5.16. Because the execution of both implementations of O_1 does not necessarily require the same time, the selector must also multiplex the *DONE* signal.

The object interface hides the execution of polymorphic behavior; that is, for the sender of a message which is handled by O_1 it must be irrelevant which implementation of this operation is executed. Thus both implementations of O_1 share the same data interface. If both behaviors of the operations change data in their data interface, the implementations must be decoupled from the actual data by appropriate logic. If no other output enable was specified in the model the *DONE* signal is used.

With a growing number of polymorphic operations the implementation requires a reasonable amount of logic resources. However, the support of polymorphism also provides significant advantages. If the implementation supports the object-oriented features the designer has more freedom for the system specification. Moreover, because object-orientation means implementing the differences between classes, the direct implementation of class hierarchies can help to reduce the amount of logic resources required. In each class of the hierarchy only the new and overridden features of the class in comparison to its superclasses must be realized. Also the probability that an object of a required type is contained in the current bitstream is raised because there are virtually more objects of different types. Hence the overall number of reconfigurations may drop. In experiments we have shown that this approach can improve the overall performance by orders of magnitude [174]. The full implementation of class hierarchies is only advantageous however, if the classes in the hierarchy are actually instantiated by the application.

The implementation of attributes and parameters is straightforward; they are mapped to a storage component of an appropriate width. If the implementation platform contains components with appropriate interfaces (`Local Access`, `External Access`), the resources modeled are used to generate the storage components. For performance and synchronization reasons the data interface of publicly visible objects is located in the register file.

Objects are connected to the data interface and other objects with direct connections or buses. This raises a significant synchronisation problem because multiple objects or operations may access features concurrently. For the publicly visible objects the software proxy objects synchronize concurrent accesses by sequentializing them. Concurrent accesses of proxy objects and their hardware counterparts are decoupled by the dual ported architecture of the register file. Potentially concurrent accesses to an element in the same hardware component are guarded with arbiters.

If the implementation uses an implementation type or operation which implements the `Device Connect` interface, the signals of this interface are routed to the top level VHDL module. This mechanism is used to include peripherals into the generated designs.

EXAMPLE 5.13 *Figure 5.17 continues the implementation of the AudioPaKCoder example. As the figure suggests, this implementation is very simple because the* `AudioPaKCoder` *neither implements polymorphic behavior nor does it access other hardware objects; hence intra-device synchronization problems do not arise. The data interface of the object is realized in the register file as presented in Example 5.12. Notably, the interface to access the sample array has been transformed into a WISHBONE-like bus interface [197]. The transformed interface was already introduced in the implementation model (see Example 5.6).*

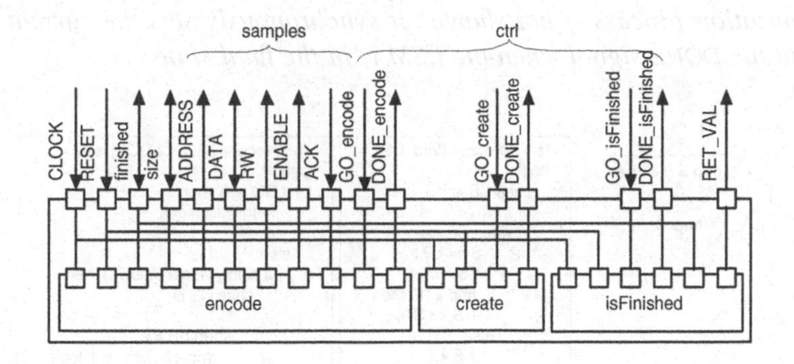

Figure 5.17. Object Implementation AudioPaK Example

5.4.6 Hardware Implementation of Behavior

The behavior of classes and operations is implemented according to the FSMD model (finite state machine with datapath) as Moore FSM [66]. This model is especially suitable for control oriented applications, and fits the message based computing paradigm of the object based model of computation.

Each behavior is constructed as controller with an attached datapath. The datapath performs the computations of the behavior, the evaluation of the conditions which control the datapath, and the components that store the inputs, outputs, and intermediate results. For the realization of the computation operations and buffers the resource services that have been allocated during design space exploration are used. The results of conditions are inputs of the controller.

Scheduling of the datapath is performed and assigned to the behavior during design space exploration. During implementation this schedule is then actually realized. For each PE the global scheduling policy is specified in the deployment platform model. For a better control of the implementation the designer may also specify a local, operation-specific policy in the implementation model.

The controller is realized as FSM. Each of the operations of the datapath is associated with a number of states of the FSM. Operations which require at most one clock cycle are associated with one FSM state. Multi-cycle operations are associated with a number of consecutive states. Independent operations and operations that execute at most one clock cycle may be chained to execute back to back in the same cycle. State transitions are performed synchronously.

EXAMPLE 5.14 *Figure 5.18 shows the realization of the first loop of the behavior of the operation* `AudioPaKCoder::encode`*. Owing to the lack of* *space, the computations in the loop are not shown. The FSM* [2] *on the left side is decomposed into a controller, a datapath, and a synchronization process. The loop is executed as long as the loop counter* `i` *is less than* `size`*. The synchronization process is not shown. It synchronously sets the current FSM state and the* DONE *signal when the FSM is in the final state.*

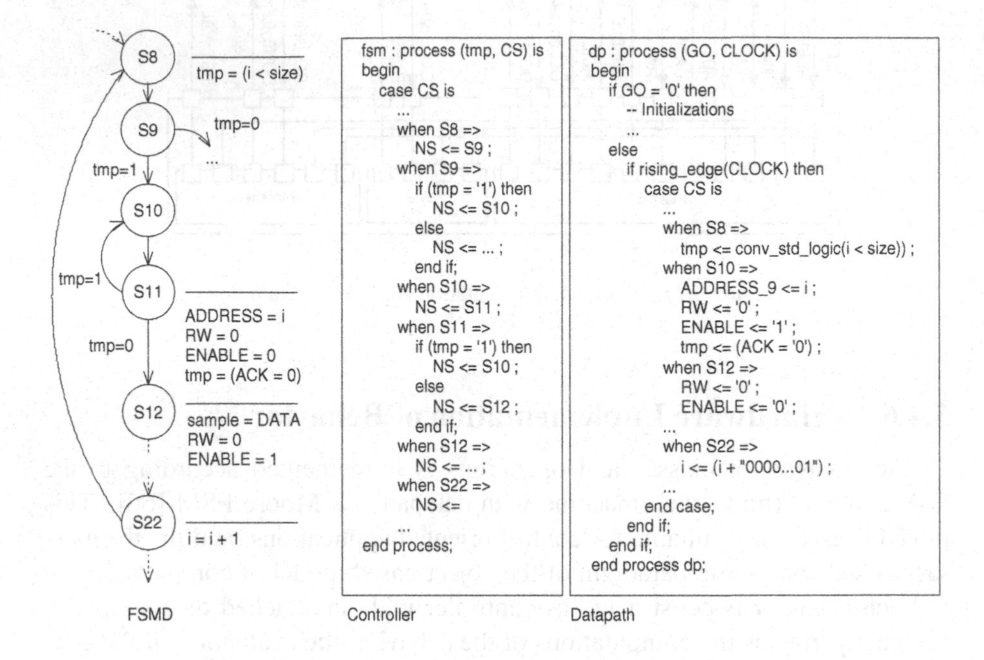

Figure 5.18. VHDL Loop Implementation Example

In contrast to software implementations each behavior is realized for each hardware object, that is, the implementation of a behavior is not shared by objects which provide the same behavior. Hence no synchronization of concurrent executions of the same behavior in different objects is required.

To avoid write contention the datapath keeps local copies of attributes and inout parameters. All modifications of the datapath are performed on the copies. If no explicit output enable is specified in the model, they are synchronized back at the end of computation and when the behavior executes a message transfer.

As shown in the previous example, array accesses in the model are transformed to explicit bus transfers. Similar transformations are performed to ac-

[2] We use the FSM notation here instead of UML State Machines in order to emphasize the relation to the VHDL implementation. Also the currently implemented FSMs have less powerful semantics as UML State Machines.

complish message exchange between objects. Message transfers between operations of the same object are commonly inlined by the model compiler. This requires more implementation resources but again minimizes data contention.

5.5 Experimental Results

The design model from our permanent AudioPaK example was automatically implemented with the MOCCA compiler using C++ and VHDL-RTL implementations platforms. The key concepts of the implementation have already been presented in the previous sections. The overall compilation/synthesis time of the design model into the final hardware/software modules takes approximately 5 to 10 minutes, depending on the degree of optimization.

The coder was tested on a hardware platform comprising a Pentium IV processor running at 2.4 GHz (master) and a Xilinx Virtex-II FPGA with approximately 3 Million gate equivalents running at 100 MHz (slave). Master and slave are connected with a 33 MHz PCI bus. The slave implements the AudioPaK coder objects, and the master is responsible for the filtering and distribution of the audio information to the coder objects and the audio clients in a network.

The current scarcely optimized FPGA implementation of one coder object requires about 1800 slices, which corresponds to 12% of the available area. Ad-

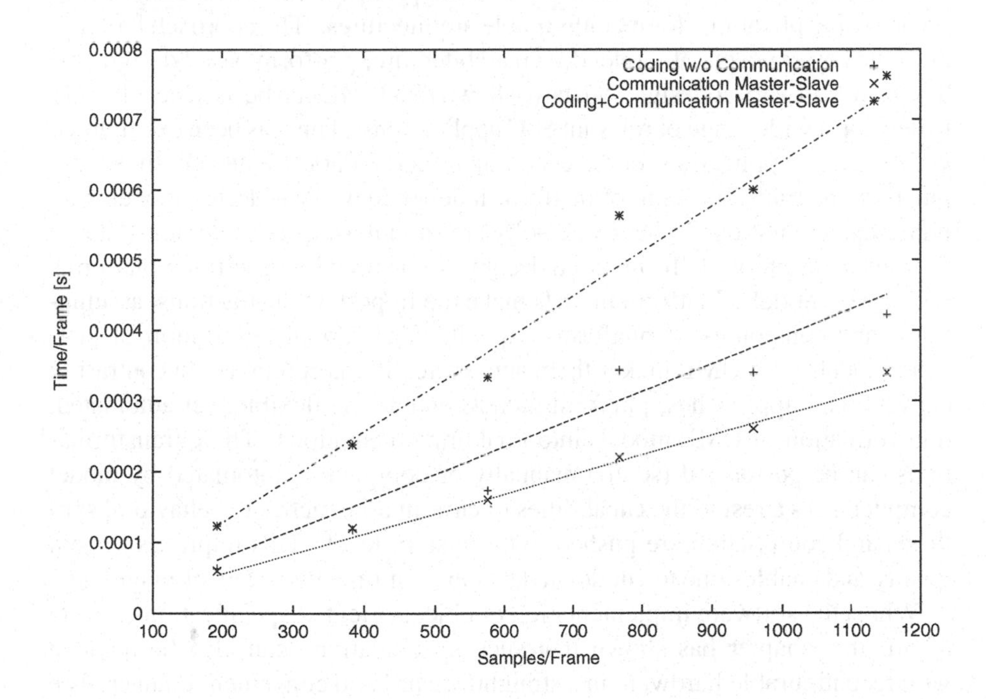

Figure 5.19. AudioPaKCoder Coding and Communication Effort

ditionally two 16Kbit BRAMs are used to store the sample array. The majority of the resources is consumed by the encode operation. The resources required for the PCI bridge are neglectable.

The FSM generated for the encode operation consists of 214 states, which results in a complex controller. Despite the relatively low arithmetic complexity of the encoding algorithm, its implementation is rather expensive. The algorithm comprises six loops realizing the intra-channel decorrelation, computation of the encoding parameters and the actual coding of the audio samples. The algorithm potential for parallelization is quite small.

In Figure 5.19 the performance characteristics of the coder implementation is shown. The coder implementation is quite efficient. Depending on the frame size, between 15 and 20 audio channels can be encoded concurrently with one coder object at 96KHz sampling rate. To avoid contention on the objects, we found it useful to implement up to five coder objects that are executed in 'round robin' order. The time to transfer the samples from/to the hardware may be dropped in the future by connecting the coder directly to the audio input devices.

5.6 Conclusions

In this chapter we have presented a novel approach to the UML-based development of applications for reconfigurable architectures. The approach incorporates the key concepts of model driven architecture, platform based design, and hardware/software codesign. We have shown that UML can be used beneficially to develop a wide range of relevant SoC applications. This has been exemplified with a simple application for the encoding of audio data streams. With this application the transformation of platform-independent UML design models into final implementations of hardware/software modules has been demonstrated.

The paradigm of platform based design can be used by specifying platforms with UML models. Platform models make the important abstractions, assumptions, and constraints of platforms explicit. The formal description of platform models with UML makes them automatically interpretable. In contrast to conventional approaches, platform models enable the flexible, yet automated, transformation of UML models into final implementations. These transformations can be performed (semi-) manually or completely automated by model compilers. As a result, the capabilities of current approaches to behavioral synthesis and compilation are pushed to the system level. This improves system quality and enables one to cut down development time by orders of magnitude.

Whilst the software implementation of object-oriented specifications is state of art, this chapter has shown that such specifications can also be realized with reconfigurable hardware in a straightforward and convenient manner. The paradigm of message based computation can be used to generate very efficient and scalable hardware implementations. The development of highly parallel applications is encouraged. Advanced model compilers for SoC can automatically implement heterogeneous multiprocessor solutions from UML models.

The capabilities of our approach have been demonstrated with a real world design of a coder for high quality audio streams. With this example it has been shown that the approach provides significant gains in system quality and development efficiency. Especially, the short implementation time motivates the exploration of different design alternatives in order to improve the overall quality and to reduce costs.

Chapter 6

A Methodology for Bridging the Gap between UML and Codesign

Ananda Shankar Basu,[1] Marcello Lajolo,[1] Mauro Prevostini[2]

[1] *NEC Laboratories America*
Princeton, NJ, USA

[2] *ALaRI, University of Lugano*
Lugano, Switzerland

Abstract The Unified Modeling Language (UML) is getting more popular among system designers due to the need to raise the level of abstraction in system specifications. We present here a methodology that integrates UML specifications with a hardware/software codesign platform. This work aims to give a contribution toward SoC Design Automation starting from system level specification down to hardware/software partitioning and integration.

6.1 Introduction

With the increasing design complexity and the reduction of the time to market windows, the design of electronic systems has become a challenging task to be handled by traditional methodologies. Embedded systems design in comparison to traditional software development requires not only to verify the functional correctness, but also to satisfy tight performance and cost constraints. Hence, new methodologies are needed to improve design productivity and derive high performance low cost implementations keeping in mind the reuse of pre-designed components.

The software community, after several years of work, converged on a set of notations for developing specifications of object oriented systems known as the Unified Modeling Language or UML [178] that has been very successful as a visual way for describing software. However, UML is not limited to software modeling and the development of UML 2.0 has been undertaken with the express

G. Martin and W. Müller (eds.), UML for SOC Design, 119–146.

intention of producing a language that has benefits for a much wider audience than just software developers, including the world of systems engineering.

In this work, we present an integration of a UML based modeling methodology with a C based design technology called ACES (Application to C to Exploration to System LSI) [109] that leverages on high level synthesis and coverification tools and aims to assist the designer in the hardware/software partitioning and architecture selection phases. ACES has the unique advantage with respect to all similar approaches to be able to leverage off the strengths of two key pieces in NEC's C based design flow [226]: CYBER, a behavioral hardware synthesis tool and CLASSMATE, a hardware/software coverification tool. UML complements ACES with an object oriented modeling language with both graphical and textual notations, organized in a set of diagrams, each diagram capturing a different aspect, or level of abstraction, of the system. The result is a unified design flow from system specification down to system implementation.

This chapter is organized as follows. Section 6.2 talks about the state of the art in electronic system level (ESL) design and focuses on our main contributions. Section 6.3 describes the ACES codesign flow, which is an integral part of our methodology. Section 6.4 talks about the hardware oriented modeling aspects of UML. Section 6.5 describes how the model can be verified in the UML environment. Section 6.6 talks about the link between UML specifications and the codesign environment. Section 6.7 presents our conclusions.

6.2 State of the Art and Contribution

As the complexity of systems increases, so does the importance of good specification and modeling techniques. Many factors contribute to the success of a project, and certainly one we cannot do without is a rigorous modeling language standard (see, e.g., [85, 142]). Introduced in recent years,UML [178] is now widely used, historically for requirements specification and for the design of complex software systems and since at least a couple of years also for hardware modeling and for embedded systems design. Although UML has a lot of advantages, is still not fully reliable for hardware description, especially for event semantics [13, 110]. This lack of semantics for hardware modeling exists because UML was originally conceived by the software development community. The Object Management Group (OMG) [144] is at the moment assessing it in order to define standard semantics able to improve hardware description modeling. These new standards have been recently adopted by OMG through UML 2.0 [217].

On the other hand, Electronic System Level (ESL) design has been a hot area for Electronic Design Automation (EDA) vendors and startups in particular, but there are so many entries now that marketplace confusion is more likely than widespread adoption. ESL point tools are many, but flows that can go from concept to implementation are few. For example, commercial hardware and software coverification tools from companies such as Mentor Graphics, CoWare, VAST, Virtio and Axys can provide fast instruction set simulators linked to various hardware simulators. They mainly focus on the functional and performance modeling problem for software dominated embedded systems, although they do not address the issues of high level hardware modeling and refinement. The main limitation of these tools is that they often require to model the hardware at the RT level and even though recently some of these vendors have started to offer the possibility to perform a mixed C/RTL coverification (e.g. C Bridge from Mentor Graphics), none of them offers yet an automated behavioral synthesis path from behavioral specifications.

An emerging area is also the one of coprocessor synthesis [133, 206, 44], where the main idea is to combine the software compilation and the hardware synthesis technologies to provide a system that allows designers to explore and implement their designs directly from descriptions written in algorithmic C. The main limitation of this approach is that it is based on the assumption that the designer has already been able to come up with a feasible hardware/software partitioning for the entire design and the coprocessor synthesizer can then provide the possibility to perform some software acceleration by offloading compute intensive algorithms from the CPU to dedicated hardware. Although very useful, tools of this type can only provide a partial support to a complete SoC design flow because it is well known that many decisions regarding the efficiency (performance, power, area etc.) of the system have largely been fixed by the time a designer commits to a particular architecture.

Alternative and complementary methodologies and solutions must hence be provided in order to help the designer during the initial phases of the design process when coarse hardware/software partitioning tradeoffs have to be analyzed. Our work is an attempt to try to fill this gap by proposing a practical integration between a UML based modeling methodology and an existing hardware/software codesign technology.

6.3 The ACES Codesign Flow

The overall flow presented in this chapter is shown in Figure 6.1. Our proposed methodology starts with the UML specification of the system, followed by exploration of the UML database for extraction of functional and structural information. This is followed by an interactive process performed through a web based interface that allows to capture UML specifications and design constraints provided by the designer, like architectural specifications and hardware/software partitioning, and export the entire structure of the design into the ACES [109] codesign environment.

In ACES the system is described at the behavioral level as a network of components that can communicate by both means of events as well as shared variables. A web based interface acts as an intermediate layer between UML and codesign through which the user can drive the codesign process by performing the important tasks of component and communication mapping. A library of precharacterized architectural templates is provided in order to allow the designer to explore different design solutions.

The following sections describe in detail the various phases in this design flow.

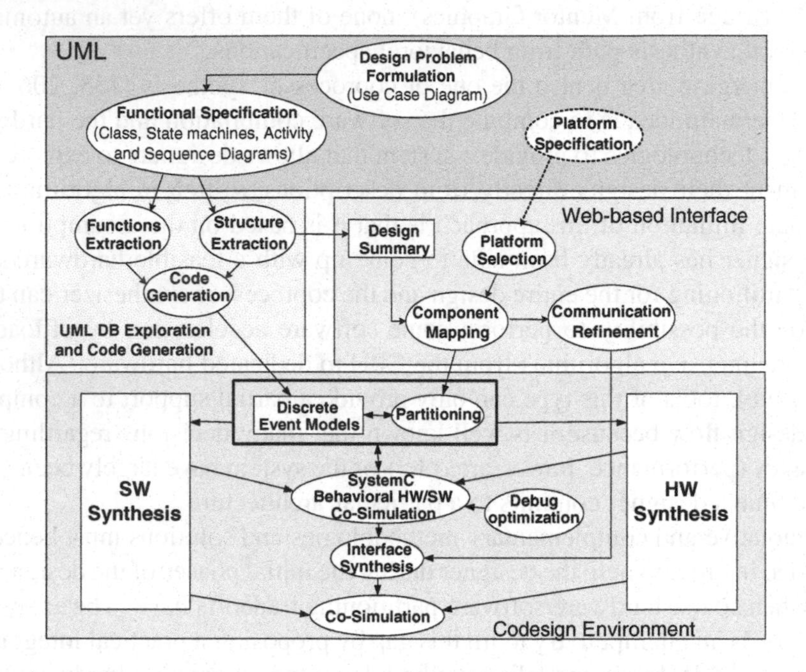

Figure 6.1. Design Flow

6.3.1 UML Specifications

UML is an object oriented modeling language that consists of graphical and textual notations, organized in a set of diagrams, each diagram capturing a different aspect or level of abstraction of the system [178]. After getting the requirements specification of the system to be designed, the first step is to capture the functionality of the system as a whole using Use Case Diagrams. In the second step, the functionality is decomposed into components within Classes describing the SoC's structure and State Machines, Activity and Sequence Diagrams, describing the SoC's behavior. Constraints (i.e. performance) are captured using Stereotypes, which are simple extension mechanisms of UML, and propagated and budgeted to the components. In the following step, the model is simulated in the UML environment in order to check whether the functional behavior of the system matches the original specifications.

For a first analysis of a possible integration between UML and codesign, we have started by considering a UML specification flow in which first an Object Model Diagram (OMD) is defined to capture the structural decomposition into interacting components. An OMD contains two sets of classes: the ones whose behavior could potentially be implemented either in hardware or in software and others that do not have to enter in the codesign flow, for example, testbenches and strictly software oriented components. The first set of classes are distinguished by a specific set of UML stereotypes and additionally they are also used to differentiate between the types of behavioral specification associated with a particular class. For example, the stereotype *Partitionable_StateMachine* is used for classes with state machine behavior and *Partitionable_Text* is used for classes with textual specifications. We will talk in more detail about this item in Section 6.4. Communication among objects of classes can be specified through links connecting ports of different objects. Ports are stereotyped as *output* or *input* which allows for semantic verification of connections between ports during the structural information exportation process. Visually, interfaces among classes are described by means of ports and connectors (see [189]). The behavior of the classes participating in the codesign process can be specified graphically using state machines as well as textually in the form of behavioral SystemC code, attached as a description to a class.

As a next step, the UML Functional Specification must be translated into ACES Discrete Event Models to conjugate the convenience of using the graphical UML Platform interface for specification with the possibility to use the analysis and synthesis tools available using the ACES codesign methodology. We have also proposed a possible way to specify the architectural platform in which system modules would be deployed onto different architectural components using the UML Deploymemt diagrams.

6.3.2 System Level API

We provide a specific API, basically an extended UML library, in order to allow the user to describe the type of communication that he wants to be performed. The API combines transaction level modeling for the hardware interface and OS and device driver levels for the software interface into a unified semantic. The objective is to provide designers with a minimal set of high level primitives that can be used to abstract and specify the behavior of the system. The proposed generic API for design specification is presented in Table 6.1. It is based on the POSIX [215] standard, a well defined and accepted programming interface for Operating Systems. The API is divided in four parts:

Table 6.1. The API Functions

Process Management	`process_create(id, param, func, arg)`
	`process_delete(id)`
	`process_suspend(id)`
	`process_resume(id)`

Communication	`port_send(port, data, mode)`
	`port_receive(port, mode)`
	`shared_mem_read(mem, offset, mode)`
	`shared_mem_write(mem, offset, data, mode)`

Synchronization	`mutex_lock(mutex)`
	`mutex_unlock(mutex)`
	`sema_wait(sem)`
	`sema_post(sem)`
	`cond_var_wait(var, mutex)`
	`cond_var_signal(var)`
	`cond_var_broadcast(var)`

Timing	`time_wait(time)`
	`process_join(id)`
	`mutex_lock_tmo(mutex, time)`
	`sema_wait_tmo(sem, time)`
	`cond_var_wait_tmo(var, mutex, time)`

Process Management, Communication, Synchronization and Timing. Process management includes functions to control process creation and execution. The Communication part encompasses shared memory and message passing based communication, both blocking and nonblocking style. Synchronization includes primitives for process synchronization, like mutexes, semaphores and condition variables. Finally, the Timing section allows some control over the timing behavior of the system, providing a timed wait and controlling timeouts for blocking operations.

The API is thought to be integrable with any system level specification language, for instance, SystemC. The range of specification styles possible to target with the API is very broad. Hardware oriented specifications might use bit manipulation and low level constructs more intensively, while software oriented specifications could use pointers, memory allocation and stack manipulation more frequently. Nevertheless, the API we propose is neutral and can accommodate either style.

In the Process Management section of the API, four functions are defined. The function `process_create` is used to instantiate and start the execution of a new process. The function `func` is the entry point of the process. Note that the actual code of the process, be it hardware or software, is already available. The API function will create a new context for the new process and start executing the initial function. Also note that in case of hardware processes, if more than one process shares the same hardware implementation, there is a need to synthesize a scheduler within the hardware implementation, so that time sharing of the hardware is possible. `Process_delete` stops and removes a process from the scheduler list forever, freeing all the resources that were held by that process. Finally, `process_suspend` and `process_resume` are used to stop and resume the execution of a process, respectively. A process is suspended by a `process_suspend` call, and stays suspended until some other process executes `process_resume` for that specific process.

Two different communication models are supported in the API, message passing and shared memory. Message passing is abstracted by the concepts of ports, and provides the primitives `port_send` and `port_receive` to implement the communication. Blocking and nonblocking styles are supported, and are specified by the designer through the argument `mode`. A blocking send blocks the sender until the receiver reads the message. Similarly, a blocking receive blocks the receiver until a message is available in the corresponding port. Shared memory communication is modeled with the `shared_mem_read` and `shared_mem_write` primitives. Here, two styles are also possible, synchronous and asynchronous, specified in the parameter `mode`.

In the Synchronization section, three different synchronization mechanisms are defined by the API: mutexes, semaphores and condition variables. A call to `sema_wait` will block the calling process if the semaphore value is zero, meaning that none of the shared resources are available, while a call to `sema_post` increments the value of the semaphore, and unblocks a possibly waiting process. Mutexes are similar to binary semaphores, i.e., semaphores initialized with the value of one. The process calling `mutex_lock` will block in case the mutex value is zero, and `mutex_unlock` will set the mutex value to one, allowing one of the possibly waiting processes to continue. Finally, condition variables allow processes to wait for some event or condition to happen. The process calling `cond_var_wait` will block until the condition is met and the corresponding `cond_var_signal` is invoked. Alternatively, `cond_var_broadcast` can be used to signal an event when multiple processes should resume execution as a result of one event.

Finally, the Timing section allows the specification of the timing behavior of processes. Processes can wait for a fixed amount of time using the API called `time_wait`. The waiting time is provided in the parameter `time`. Additionally, it is also possible to specify timeouts for each of the blocking synchronization primitives, with `sema_wait_tmo`, `mutex_lock_tmo` and `cond_var_wait_tmo`.

6.3.3 Interface Synthesis

When the input design description contains communication primitives from the System Level API, there is a need to synthesize the communication interface between the processes. Depending on the design partitioning, the interface will need to connect two hardware modules, two software modules, or a hardware and a software module. This phase is controlled through a web based interface that acts as an intermediate layer between UML and codesign and that will be presented in detail in Section 6.6.4 in the context of a real application.

In this section, we show examples of custom interface synthesis for different partitions. We refer to the process sending data as the producer, and the processor receiving data as the consumer.

Hardware to Hardware Communication. In the case where two processes that communicate through ports are mapped to a hardware implementation, there are different alternatives for interface synthesis. However, since this is a hardware to hardware communication, it is not necessary to generate RTOS code or software to handle this specific communication.

One possible architecture for a port based hardware to hardware communication is shown in Figure 6.2. In this case, there is a direct data connection between producer and consumer. Additionally, control lines are synthesized according to the API usage. If the port is ever used for a blocking send, then an acknowledgement line from the consumer to the producer is necessary. Therefore, the producer is suspended until it receives an acknowledgement from the consumer in case of a blocking communication. For communications with multiple consumers, the producer waits for the acknowledgement of all consumers. This behavior is implemented with a logic OR of the individual acknowledges of the consumers, as shown in Figure 6.2. Similarly, an event line is added from the producer to each consumer for the case when blocking receives are specified. Since the event and acknowledge control signals are only synthesized when needed, they are shown with dashed lines in Figure 6.2.

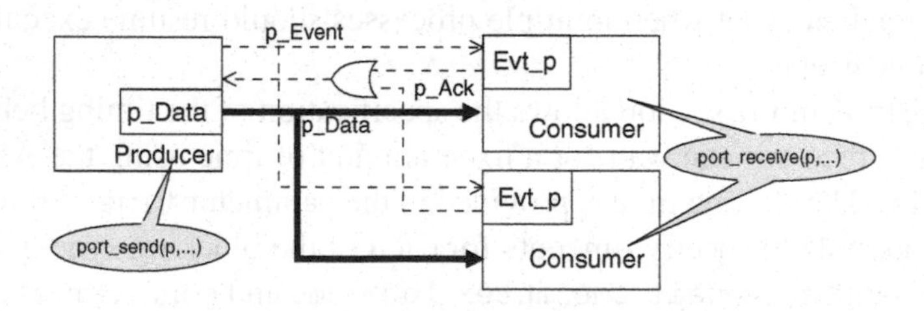

Figure 6.2. Interface Synthesis for HW to HW Communication

Other architectures are also possible from the same System Level API. For instance, it is possible to generate a Transaction Level Model with AMBA bus transactions for each port primitive. In this case, the `port_send` and `port_receive` primitives are replaced by a set of calls to the AMBA Transaction Level API [6].

Software to Software Communication. When two software processes are mapped to the same processor, the interface synthesis is simpler. Our framework will generate a software data structure in memory, shared between the processes, that will keep the data along with event and acknowledge control signals. All the producer has to do is to update two memory locations, with data and event signaling (in case of blocking receives), while the consumer will read the data memory and update the acknowledge bit of the same port. Figure 6.3 shows the interaction between the processes.

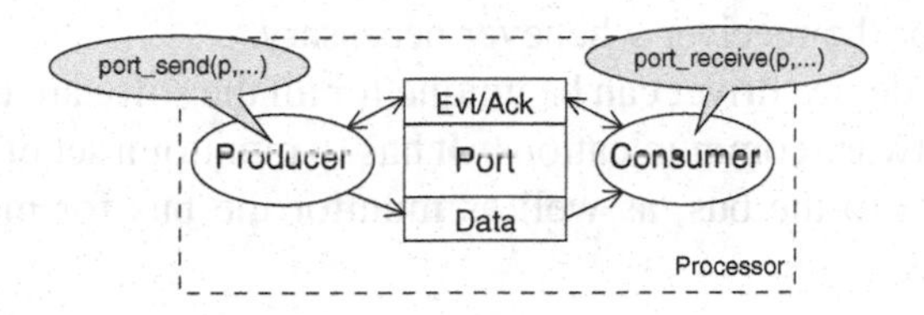

Figure 6.3. Interface Synthesis for SW to SW Communication

Hardware to Software Communication. Hardware to software communications can be implemented by either interrupts or polling, using memory mapped addresses in the latter case. In both cases, we will need some RTOS support in order to coordinate the processes. One possible solution is shown in Figure 6.4. Our framework will generate a bus adaptation layer for the hardware module, so that it can send and receive data from the bus. In the case of a memory mapped communication, a device driver is also generated and runs inside the processor, monitoring the bus for activity in the memory mapped region. The device driver is responsible for transferring data from the bus to the processor memory, to an equivalent port structure as the one shown in Fig-

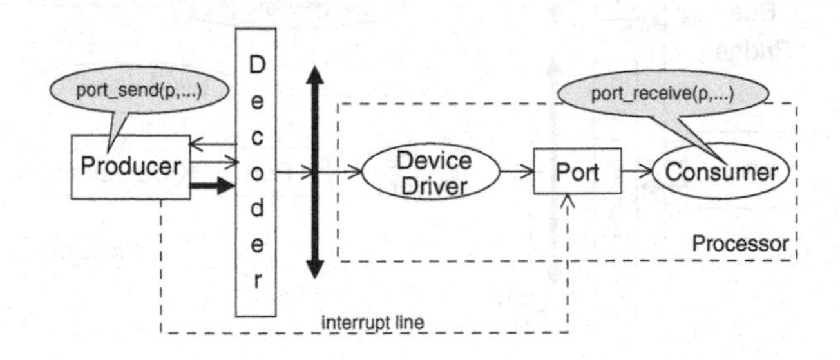

Figure 6.4. Interface Synthesis for HW to SW Communication

ure 6.3. The software process will access the port data structure as it did in the software to software case, retrieving data and updating event flags. If instead an interrupt based communication is specified, then an Interrupt Service Routine (ISR) needs to be synthesized. The ISR will be responsible for receiving the event signaling from the producer. In the interrupt based communication, the actual data is still transferred through a memory mapped location to the port structure.

Software to Hardware Communication. In software to hardware communications, the producer is running in a processor, communicating with a hardware module. In our model, this kind of communication is always memory mapped. The producer will update a *port* data structure, and a device driver propagates data and events to and from the bus. Events and acknowledge signals are generated for the receiver whenever necessary.

Note that the device driver can be unique for all the software to hardware and hardware to software communications. It has to monitor a set of software ports, transferring data to the bus, as well as monitor the bus for memory mapped communications.

Multiprocessor Communication. Finally, in case the processes are mapped to different processors, with different buses, a bridge will also be synthesized. Figure 6.5 shows the proposed architecture. In this scenario, the producer runs on processor 1, connected to CPU1 Bus, while the consumer runs on processor 2, connected to CPU2 Bus. The producer will see the bridge as the consumer. Meanwhile, the consumer will see the bridge as the producer. Both processes will see a hardware to software communication, and the port will be accessed through a memory mapped address. In addition to the bridge, device driver code

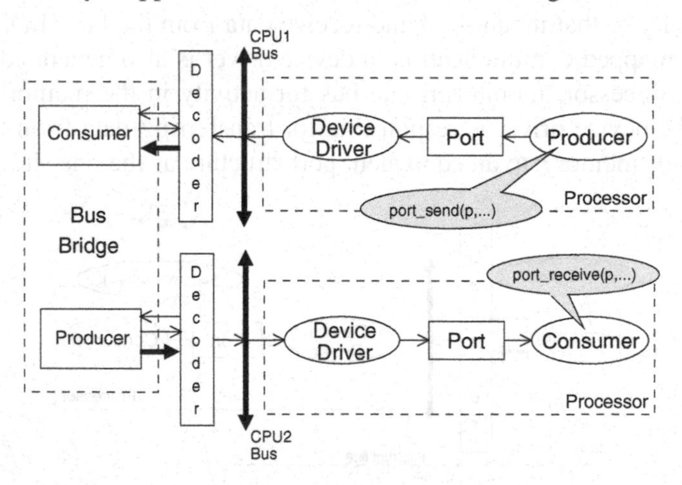

Figure 6.5. Interface Synthesis for Multiprocessor Communication

is synthesized for both processors, linking the software process to the RTOS and to the bridge hardware.

For shared memory communication, two different architectures are possible, depending on synchronous or asynchronous communication. In the synchronous mode, a locking structure is generated for each shared memory, so that access is granted exclusively for each process. Every memory access has to obtain the lock first. In the asynchronous mode, only the memory is synthesized. The locking mechanism is implicit in the API call for shared memory access. Every shared memory will be directly connected to the system bus, accessible by the CPU. Additionally, a dedicated memory port will be available for each hardware module accessing the memory, so that using the bus is not necessary while accessing shared data. Therefore, there is less contention and higher parallelism in the implementation.

RTOS Synthesis. In addition to communication interface synthesis, the generation of RTOS support is required. In this case, our System Level API has to be mapped to OS specific resources, adapting the generic API to the functionality available in the target RTOS. Since our API is based on POSIX, the mapping is trivial when targeting a POSIX compliant OS, like Embedded Linux [214]. Alternatively, it is possible to target non POSIX RTOSes by mapping the API calls to the specific RTOS. That is the case with eCos [129]. Finally, the API based description can be used as input to tools that generate a customized OS infrastructure, like Polis [10] and Phantom [140].

6.3.4 Our HW/SW Codesign Environment

Input to our codesign environment is a set of modules $M_1, M_2 \ldots M_n$ that implement a design. Modules are described in SystemC extended with the proposed API functions. The SystemC modules are partitioned into hardware and software implementations. Currently, the partitioning process is manual. Once the design is partitioned, hardware, software and interfaces are synthesized. Hardware synthesis is handled by an inhouse SystemC behavioral synthesizer, that produces synthesizable RTL for each SystemC module. Software modules are generated according to the operating system support desired by the designer. At the time, our environment can generate software modules based on the POLIS framework [10], the Phantom Compiler [140] and any POSIX based operating system, like Embedded Linux [214] or eCos [129] with the POSIX adaptation layer. Software is compiled to a specific processor, which can be a NEC V850 or an ARM946. Finally, the interface is generated according to the partition and the communication style specified. We have simulators available that allow us to simulate the synthesized hardware, selected processor (cycle accurate in the case of V850 and instruction based in the case of ARM), software and communication interfaces.

6.4 Modeling Hardware Related Aspects in UML

This section deals with the hardware oriented modeling aspects of UML. In particular, it describes methods for specifying a system using the different flavors of UML diagrams, some depicting the structure and some depicting the behavior. It shows how hierarchy in hardware design can be represented at the specification level using available UML features. It also talks about our proposed enhancement of textual specifications and ways to integrate that in our codesign flow. Lastly, it speaks about the UML2.0 enhancements relevant to hardware oriented support, in particular the usefulness of timing diagrams as well as the specification of interfaces, ports and connectors. To create our model we used Rhapsody V5.2 which is the UML tool provided by I-Logix Inc.

6.4.1 Object Model Diagrams (OMD)

The OMD helps designers in modeling the structure of the system by means of classes. In our design flow we assume that each instantiated class is a functional system component. Figure 6.6 describes the top level OMD views for our example model that implements a simple matrix multiplication algorithm. It shows the block IndexControl, that controls the execution of the algorithm, a memory object and a hierarchical block MatrixMult. The OMD shows the static structure of the specified system, in particular, classes and their internal structure like the objects instantiated within them and relationships among the objects. The OMD can also show the relationships of a class with respect to other classes, such as inheritance or generalization and associations. We have used the OMD in a way to show the hierarchical view of a design, where each OMD shows the details pertinent to that hierarchy. In this way, the OMD can be used to represent hardware modeling. In our system, there are three basic partitionable objects (IndexControl, DataRetrieve, Multiplier) and a memory object. We have created two OMD's to show the hierarchical break down of the design. The top level object model diagram in Figure 6.6 shows the highest level view of the design under test. Hierarchical objects are marked with stereotype *net* while basic partitionable objects are marked either as *Partitionable_Text* or *Partitionable_StateMachine*. The top level view also shows the input and output stimuli that needs to be generated from the test benches in order to simulate the model. The hierarchical component MatrixMult is described in Figure 6.7 which shows its component classes like DataRetrieve and Multiplier. Also note that the same Memory object appears in both the OMD views to show the relationship it shares with different objects across different levels of hierarchy. All relationships between objects are specified using *links*, which are connected via *ports*. The links can specify event based communication or pure data communication based on the stereotype attached to them. Event based communication triggers a transition in a state machine (see Section 6.4.3), but pure data communications do not trigger any transitions. Another kind of link is used to specify the relation of an object to a memory. These links have an associated direction and are shown as an arrow. For example, in Figure 6.6, the link between IndexControl and Memory belongs to one of this type.

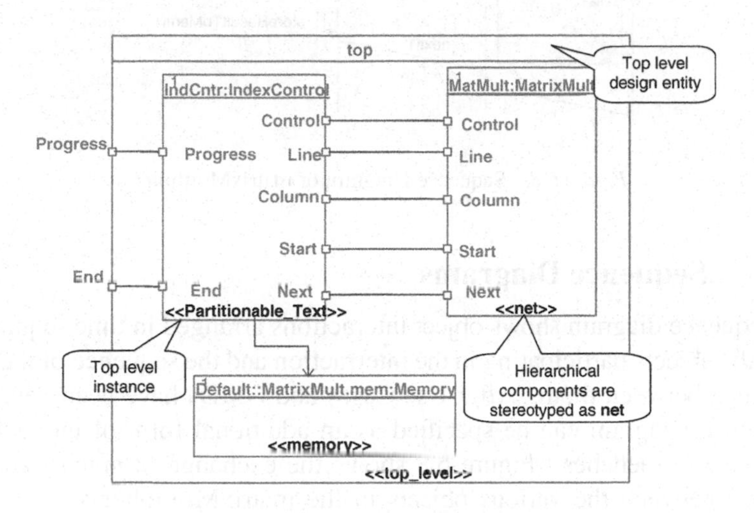

Figure 6.6. Top Level Object Model Diagram

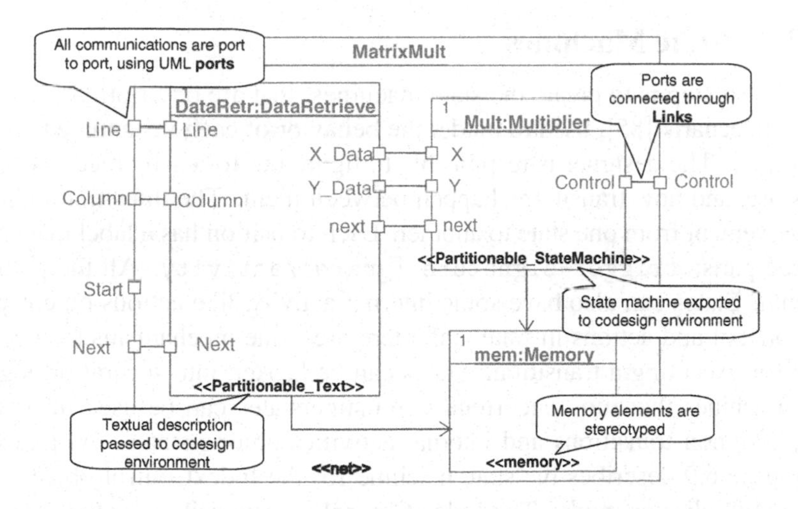

Figure 6.7. matrixMultiplier Object Model Diagram

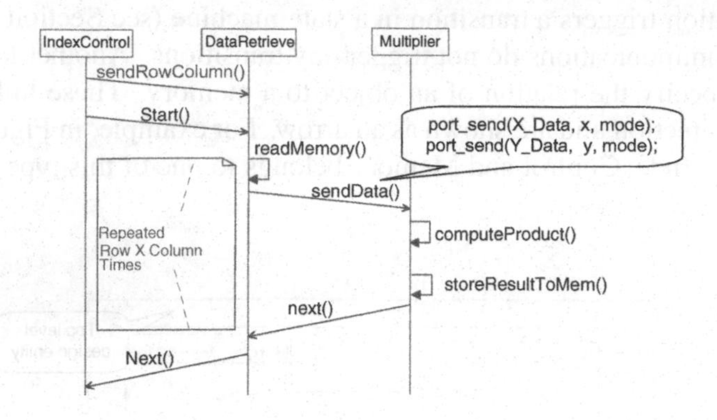

Figure 6.8. Sequence Diagram of matrixMultiplier

6.4.2 Sequence Diagrams

A sequence diagram shows object interactions arranged in time sequence. It shows the objects participating in the interraction and the sequence of messages exchanged between them. After Use Cases and OMDs have been developed, a Sequence Diagram can be specified as an additional form of interaction to help create testbenches. Figure 6.8 shows the exchange of signals and their sequence between the various objects in the matrixMultiplier system. Data communication across the objects through the function calls is shown in the figure. API calls can appear inside the functions, for example the function *sendData()* calls internally the API *port_send()*.

6.4.3 State Machines

The next step is to create the state machines, that are descriptions based on Harel statecharts [85], used to model the behavior of each instantiated class in the system. The designer is responsible to figure out for each objects what the states are, and how transitions happen between them. The transition indicates one movement from one state to another. Each transition has a label that comes in three parts: `trigger-signature [guard]/activity`. All the parts are optional. States can also have some internal activity, like actions on entry, actions on exit and actions in state, and there are some mechanisms to specify a delay for executing a transition. States can be broken into several orthogonal state machines that run concurrently and superstates can be used in order to share common transitions and internal activities among states. As an example, Figure 6.9 describes the state machine for the IndexControl object in the matrixMultiplier example. The Index Control is responsible for the execution sequence of the matrix multiplication. It is basically composed of two nested loops, that advance the current line and column of the multiplication. Current

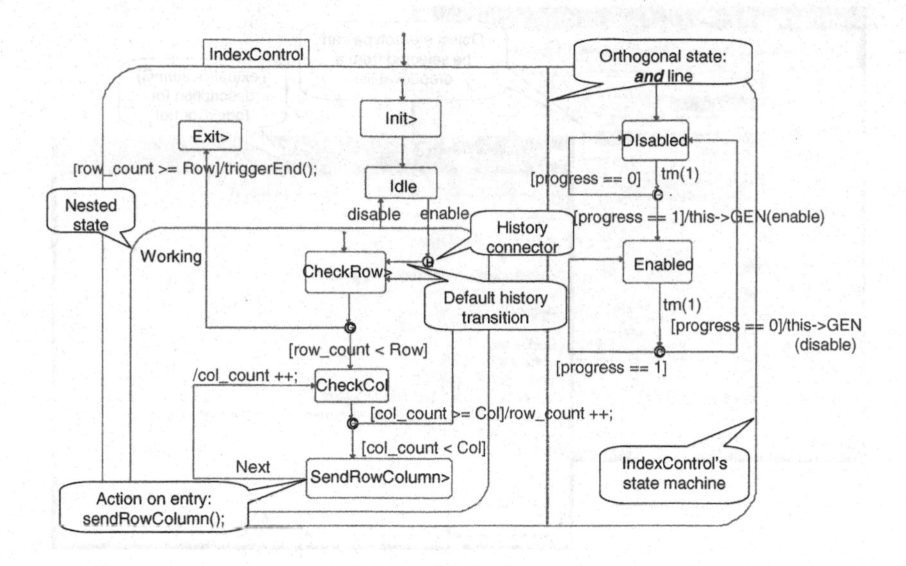

Figure 6.9. State Machine of IndexControl

line and column are communicated to the Data Retrieve module by two ports, named Line and Column.

Here is shown an AND state containing a nested state machine with a `history` connector. An AND state is an orthogonal state which represents simultaneous independent substates that an object can be in at the same time. A `history` connector stores the most recent active configuration of a state, so a transition to a history connector restores this configuration.

6.4.4 Textual Specifications

Figure 6.10 shows how we manage the textual format of the behavioral description. Through the Rhapsody tool we have to set, for each system component, which is the stereotype that specifies the type of the behavioral description. In the "Description dialog box" we edit the textual (SystemC) description of the system module. A module can potentially contain both a state machine as well as a textual description for its behavior in the form of SystemC. In the codesign phase, it will be possible to associate different instantiations of the same module to different form of specifications, i.e. some to state machines, some to textual description.

6.4.5 UML 2.0 Enhancements

UML 2.0 enhancements are not changing dramatically the modeling elements of UML 1.x. As it is said in [59]: "UML 2.0 doesn't represent a substantial

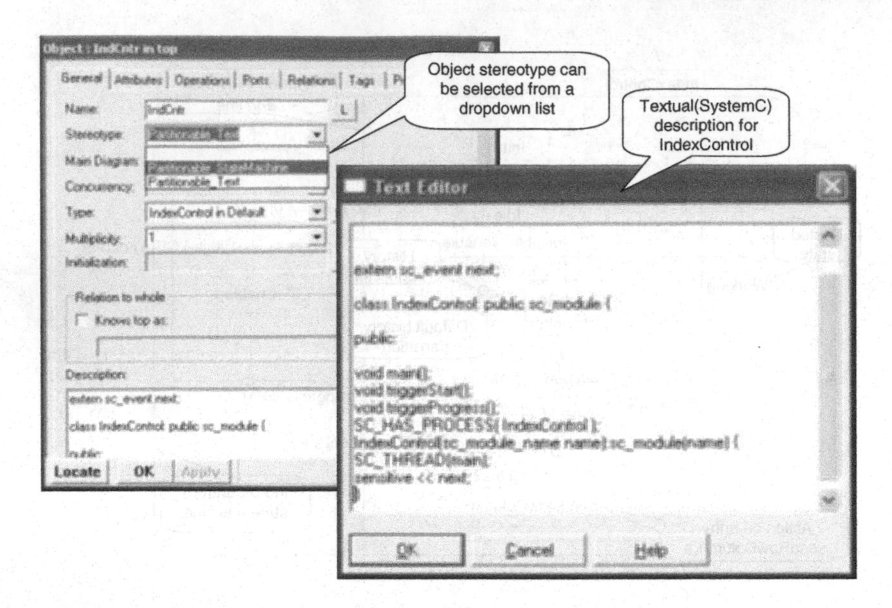

Figure 6.10. Behavioral Description: Textual Format

redefinition of the modeling elements". Most of the changes were performed in the behavioral diagrams rather than in the structural diagrams. For instance in the interaction diagrams the collaboration diagram disappeared, while the sequence diagram notation is now able to support nested diagram notation and conditional behavior. In addition there are three new interaction diagrams: timing, communication and interaction overview diagram. Among the behavioral diagrams, the activity diagram was improved significantly and now it is possible to model the concurrent behavior relying on tokens similar to Petri Nets [171].

In this section we will talk more about UML2.0 enhancements relevant to hardware modeling. In particular we will focus on structural diagrams like the Deployment diagrams. We will also spend a few words on an interaction diagram, the timing diagram, and its related elements specialized for realtime systems. In addition we will talk about the specification of interfaces, ports and connectors.

Deployment Diagrams. A Deployment diagram captures the configuration of runtime processing elements and the software component instances that reside on them. It is a graph of nodes, representing the hardware resources, and communication paths representing physical connections among the resources. A node can be a CPU or some other processing element and can have its own memory. Components represent software modules, tasks or processes that run on a node. Hence deployment diagrams specify the runtime physical architecture of a system.

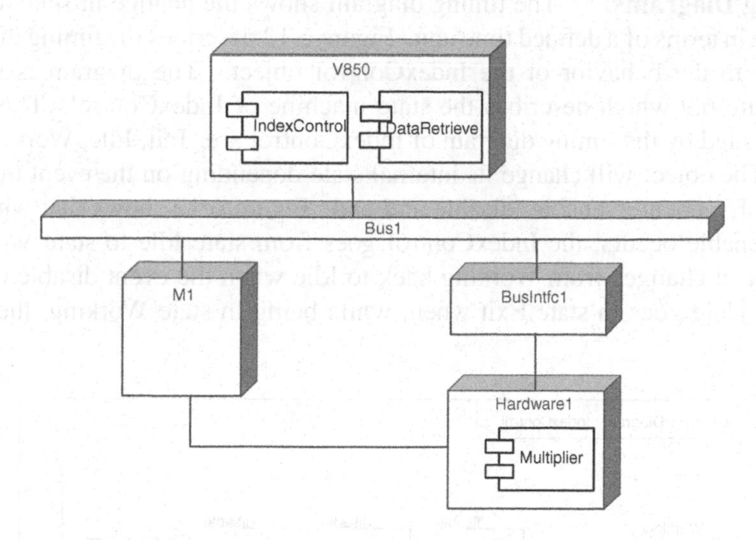

Figure 6.11. Deployment Diagram: Platform Specification and Component Mapping

Platform Specification using Deployment Diagrams. The deployment diagram can be used to specify a platform architecture in our proposed methodology. Normally the user has to select an architecture from a list of predefined platforms to be considered in the codesign phase. These predefined platforms are shown as an interconnection of hardware resources like CPU, hardware elements and memories, connected by means of buses or dedicated point to point connections. Depending on the user's choice to map a design module either to hardware or software, the modules are deployed to the corresponding elements in the architectural platform. We propose an additional stage in our design methodology where the platform can also be specified graphically by the user, making use of the UML Deployment diagram. Other than the predefined platforms, we would also provide the basic resources like the CPUs, switches, hardware elements, buses, bus bridges, etc., from which the user can select, and connect them using communication paths to build his own platform. There would be necessary checks to ensure the semantic correctness of the usage of the architectural components as well as their interconnection protocol. The design components can then be deployed to the nodes in the architectural platform. Information from the deployment diagram would then be exported to the codesign environment for further steps to cosimulation.

An example of our proposed scheme is shown in Figure 6.11. It shows the V850 platform specification and the default configuration of the modules in the matrixMultiplier application.

Timing Diagrams. The timing diagram shows the change in state along a lifeline in terms of a defined time unit. Figure 6.12 describes the timing diagram related to the behavior of the IndexControl object. The diagram is related to Figure 6.9 which describes the state machine of IndexControl. The states represented by the timing diagram of IndexControl are: Init, Idle, Working and Exit. The object will change its internal state depending on the event that will occur. Events are: enable, disable and end. Figure 6.12 shows that when the event enable occurs, the IndexControl goes from state Idle to state Working, whereas it changes from Working back to Idle when the event disable occurs. The module goes in state Exit when, while being in state Working, the event end occurs.

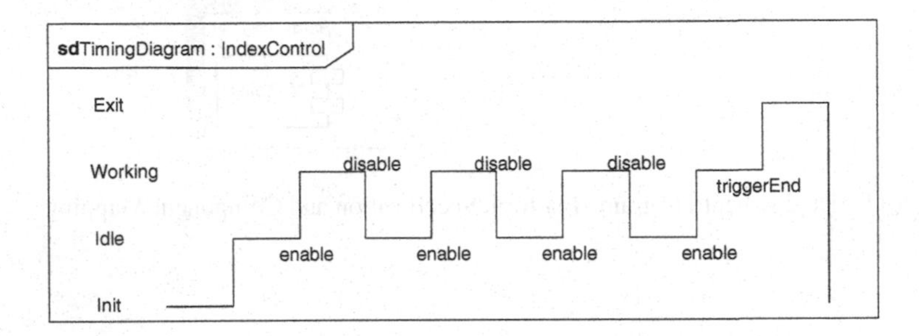

Figure 6.12. Timing Diagram for IndexControl

Timing diagrams are very helpful to specify the duration and timing constraints of realtime systems.

Interfaces, Ports and Connectors. Figure 6.13 shows how we model interfaces, ports and connectors between IndexControl and MatrixMult objects. Interfaces are specified through ports and connectors. Ports are identified by little squares on the object boundaries while connectors might be little plain circles or arcs. Circled connectors describe the provided interface (e.g the object sends a signal through this port), arc connectors describe the required interface (e.g. the object waits for a signal through this port).

Interface direction (input/output) can be shown graphically only using UML 2.0 semantics which allows to specify whether an interface requires (input) or provides (output) a service (a signal in our case study). For links, the direction cannot be specified, so we are using an input/output stereotype attached to a port in order to specify the direction. Connections to memories do not use ports, but direct links with objects.

A port is an interaction point assigned to an object and can exchange messages with other external objects or send messages to and from their parts. A port enables to specify instantiated classes independently of the environment in

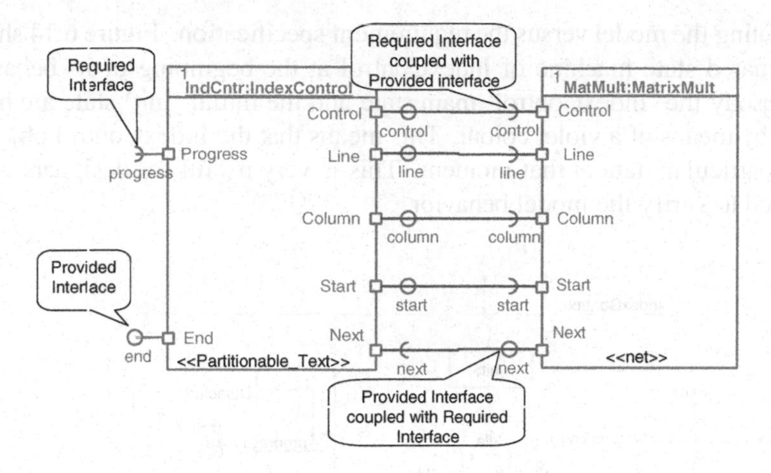

Figure 6.13. Interfaces, Ports and Connectors between IndexControl and MatrixMult

which they will be embedded. The internal part of the object can be completely isolated from the environment. In our methodology, ports are used to extract the interface signals of a module needed in its SystemC implementation (see Section 6.4.1).

6.5 Model Verification in UML

This section deals with verification aspects of UML. In particular, we discuss the event semantics in UML, and propose some UML enhancements for supporting a pure discrete event simulation that is more suitable for hardware modeling. We also describe how to take advantage of other useful UML features like animated sequence diagrams and state machines during the system verification process.

During the design phase, designers should periodically validate their UML models so that they can find bugs very early in the project. Discovering bugs in the design phase is much cheaper than in later phases. In this section we present animated diagrams, that are important features provided by UML tools, and we also discuss event semantics in UML.

6.5.1 Animated Sequence Diagrams and State Machines

The first technique we use is a particular feature provided by the UML tool Rhapsody from I-Logix [172], which allows the designer to simulate the model by animating its sequence diagrams and state machines. This allows the designer to visualize the system behavior during a specified test case and validate the model. Rhapsody also provides the possibility to compare the animated sequence diagrams with those developed during the Analysis phase. This helps in validating the model versus the requirement specification. Figure 6.14 shows the animated state machine of IndexControl at the beginning of its behavior. In Rhapsody the 'IndexControl' main state and the initial 'Init' state are highlighted by means of a violet colour. This means that the IndexControl object is in that particular state at that moment. This is very useful for designers when they need to verify the model behavior.

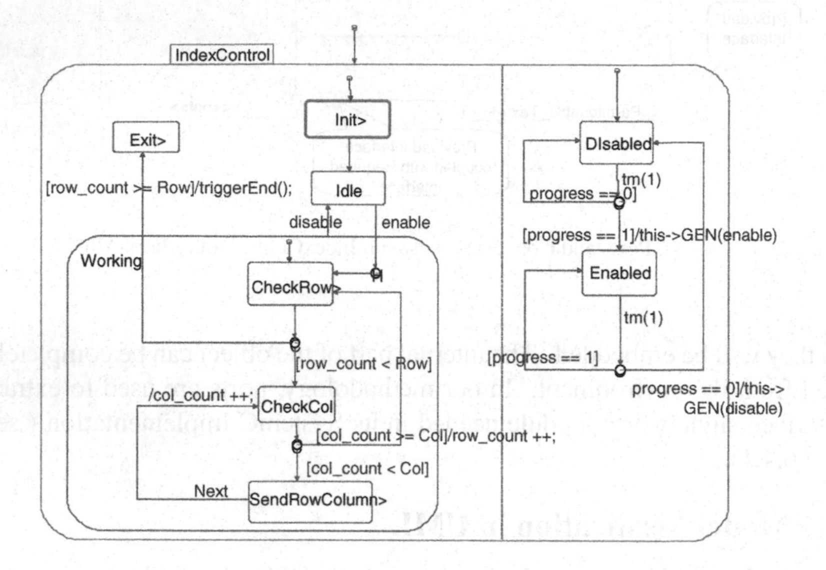

Figure 6.14. Animated State Machine

6.5.2 Event Semantics in UML

The native communication model in UML is based on asynchronous events with a single queue. In this model events are processed in the following fashion:

1. An event is created when it is sent by one object to another.
2. It is then queued on the queue of the target object thread.
3. An event that gets to the head of the queue is dispatched to the target object.
4. The event is processed by the receiving object and then deleted by the execution framework.

The main drawback of this semantics is that it is essentially SDL-like, and hence it is not adequate for hardware modeling where instead a real discrete event engine would be needed. In particular a global event queue and a support for event ordering based on their timestamps is mandatory for simulating real hardware.

UML 2.0 has tried to address this issue by allowing management of events as an event pool without defining a priori their order of dispatching. This leaves open the possibility of modeling different models of computation. Methodologies like the one presented in [77] have already shown how it is possible to extend UML very easily and efficiently in order to support new models of computation.

Unfortunately, many UML tools that are currently on the market still use the native communication model of UML and hence it is not yet possible to rely on a general solution for modeling hardware behavior in UML. For this reason, we have decided to use UML only for system specification but not for hardware/software cosimulation and validation.

6.6 Transformation from UML to Codesign

In this section we present the link between UML specifications and the ACES hardware/software codesign environment. Items that will be discussed are: UML database exploration, behavioral code generation and export of structural information necessary for the codesign environment.

6.6.1 UML Database Exploration

After the application is modeled and analyzed using the UML tool, we get a repository that contains information of the model in the internal database. We have used Rhapsody from I-Logix, Inc. as UML tool. The database generated by Rhapsody is organized as follows. The main project consists of a list of packages. A package is a mechanism to organize different project elements into groups. A package consists of the list of classes, functions, objects, events, global variables, diagrams as well as packages. Each class has a list of attributes, methods, objects instantiated within it, links between the objects and other modeling elements.

6.6.2 Code Generation

In this phase, the behavior of the modules specified in UML is converted to SystemC code in order to be imported into the ACES codesign environment. From this SystemC representation, ACES is then able to perform both hardware and software synthesis. In our proposed methodology, the behavior of a module can be specified either through a state machine as shown in Figure 6.9 or a textual specification as shown in Figure 6.10. The type of the specification can be selected from a drop down list.

When the user selects the textual specification, the code generator just copies the user specified code into the file needed by ACES. Input and output descriptions for the ports and signals would be automatically extracted by the code generator from the OMD.

On the other hand, when the user specifies the behavior through a state machine, the code generator has to explore the UML database in traversing the states of the state machine and generate the corresponding code. A sketch of the algorithm for this code generation process is shown in Figure 6.15. The algorithm needs to know the events in the model, the action/guard for the transitions, entry and exit actions in a state, in transitions to a state, out transitions from a state, etc. It is possible to browse through the entire object model and extract the relevant information from the state machine. Also UML allows the behavior to be specified using Activity Diagrams, which are very similar to State machines, and the same algorithm can be used for generation of the behavioral code.

The algorithm is called on the root state of each state machine for which code has to be generated. In every state, it first emits the code specified by the user in the action on entry portion of the state. Then it checks out transitions from the state. For the transitions triggered by events, it issues a wait statement on that event, then it emits the code specified in the action on exit portion, followed by a goto statement, the label being the target state. In case of a conditional transition, it issues an "if then else" statement with goto labels depending on the condition. It also issues the code (if any) specified in the action section of the transition. Then the algorithm is called recursively on each state reachable by the current out transition. For an **AND** state, the same code generation algorithm is called on each of the substates within the **AND** state. The behavior is also the same for a state with a nested state machine.

The output of the code generator is a list of SystemC files, each corresponding to the behavior of a specific object in the system to be considered in the codesign flow. Figure 6.16 shows a portion of a SystemC description generated for the state machine of object IndexControl corresponding to Figure 6.9. We have chosen an unstructured style for the generated code, due to its simplicity and efficiency, but many variants (e.g. nested switch, state pattern, state tables, etc.) are possible. Events are implemented as boolean terminals.

We have implemented and tested the algorithm using Rhapsody UML tool, which provides API functions that allow us to extract all required information from a UML project database. However we would like to emphasize that this code generation algorithm is very general and can be utilized also with other UML tools.

6.6.3 Exporting Structural Information

In order to start with the codesign process, the last thing we need is to extract a summary of the design, essentially a textual representation containing a list of

```
codeGenerate(state S) {
 1. If S is visited, return;
 2. Mark S as visited.
 3. Issue code specified in the action-on-entry section (This code can
    be directly copied)
 4. Get out transitions {T}  from state S;
 5. {U} = empty;
 6. for each out-transition 't' of {T} do {
      if 't' is conditional {
         issue code specified in the action-on-exit section;
         s_t = target state if condition is true;
         s_f = target state if condition is false;
         issue if-then-else with goto label as 's_t' or 's_f' depending
         on condition;
         insert 's_t', 's_f' in {U}; }
      else {
         s = target state of 't', insert 's' in {U};
         if 't' is triggered by event 'e' {
            issue wait on event 'e';     }
         issue code specified in the action-on-exit section;
         issue goto with label as 's';  }
      issue code specified in the action section of transition 't'; }
   for each 'u' in {U} do
      codeGenerate(u);
 }
```

Figure 6.15. Algorithm to Extract Code from UML State Machine

```
#include <IndexControl.h>
extern sc_int<8> mem[Row*Column]; // External memories
 SC_MODULE(IndexControl) {
 sc_in_clk clk;
 sc_in<bool> rst;
// Input terminals
sc_in<bool> Next;      // input event */
// Output terminals
sc_out<bool> End;   // output event
sc_out<sc_int<8>> row; // output data
... Omitted ...
 SC_CTOR(IndexControl) {
   SC_CTHREAD(main,clk.pos());
   watching(rst.delayed() == 0);
 }
void main(void) {
 ...
CheckRow:
        col_count=0;
        if ( row_count < Row ) {
                goto CheckCol;
        }
        else {   triggerEnd(); // Send End
                goto Init;
        }
CheckCol:  ... Omitted ...
 };
```

Figure 6.16. Code Generated for Module IndexControl

all the partitionable objects and their interconnections as well as a description of the memory object. More specifically, this phase generates the files which are necessary for ACES to use as input to proceed to cosimulation.

```
Main {
  Open UML project database
  Get list of packages
  For each package do {
    get list of defined classes
    find class from list marked as top_level
    DFS_Traverse(top level class) }
}
DFS_Traverse(class C) {
  Mark class as visited
  get list of object instantiations in class
  For each object do {
    If object is a memory instance {
      generate memory descriptions }
    else {
      generate structural descriptions
      put object's master class in DFS_List }
  }
  For each master class in DFS_List do {
    DFS_Traverse(master class) }
}
```

Figure 6.17. Pseudocode for Extracting Structural Information

In order to export the structural information to ACES, we need to traverse the design hierarchy and generate the textual descriptions. The algorithm is described in Figure 6.17. The algorithm makes a breadth first search traversal of the design hierarchy and generates the text files. In order to identify the highest level of the hierarchy, the user needs to specify a stereotype *top_level* to the top level module. Any intermediate hierarchical modules are stereotyped as *net*, whereas the leaf level modules are marked either as *Partitionable_StateMachine* or *Partitionable_Text*. Example of the text files generated for the matrixMultiplier example is shown in Figure 6.18. The generated files consists of the following:

1. A file describing the structure of the system. It consists of all the class instantiations in a hierarchical fashion, showing the inputs and outputs at each level of hierarchy and also the port connection of the instantiated classes.

2. A file describing all the signals that are necessary for connecting the objects of the system. It shows the list of signals along with their source and destination objects, and also in particular the ports of the object with which the ends are connected. Any source or destination which is in the outer hierarchy is shown as *OUT*.

3. A description of the memory objects used in the system, specifying the memory name, objects that access the memory, and other details like total size of the memory, word length and access type.

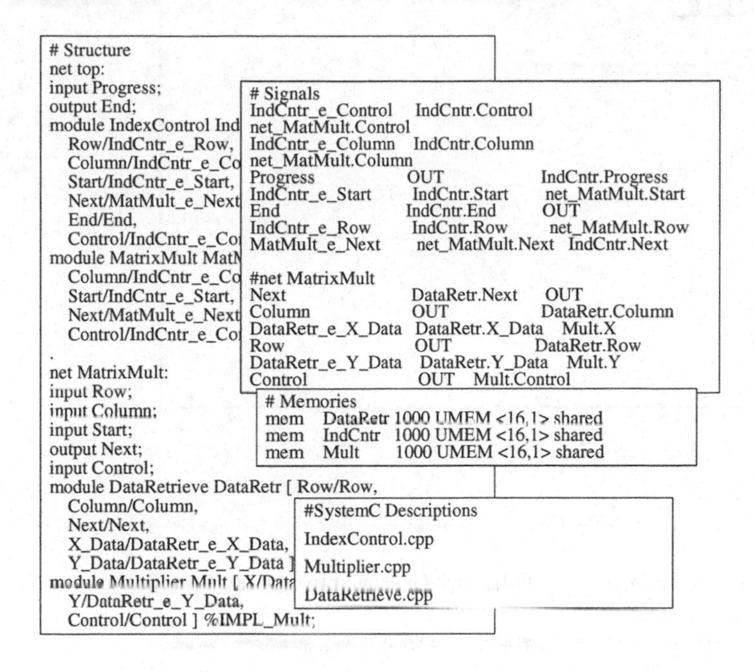

Figure 6.18. Generated Input Files for Codesign Environment

6.6.4 Web Based Interface

Figure 6.19 refers to the HTML page that is generated at the beginning of this phase. The two screen shots show the same page and respectively the top part (left side) and the bottom part (right side). This page can be opened using any web browser and is organized as follows. Starting from the top, there is a brief summary of the project containing its name and a short description. By clicking on a link, it is possible to see all the verbose reports provided by the UML tool containing all the information about the project that has been collected in the UML database. The third line is used in order to select the platform on which to implement the desired functionality. The selection is performed through a menu window where the user can pick any of the architectural templates available in a library provided with the codesign tool. An architectural template represents the platform for the system implementation and the user is responsible for selecting the platform that is best suitable for the system that he needs to implement (one or multiple CPUs, DSPs, simple or very complex bus hierarchy, etc).

The selection of the platform is directly reflected in the graphical content, presented in the middle of the page, where on the left side there is the functional

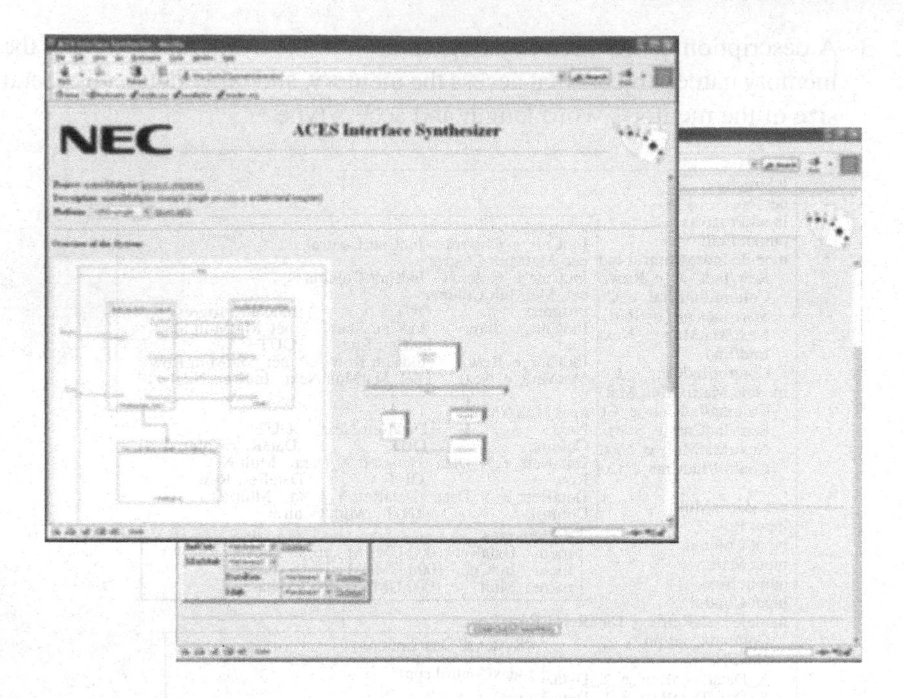

Figure 6.19. HTML Page Generated from UML Specifications

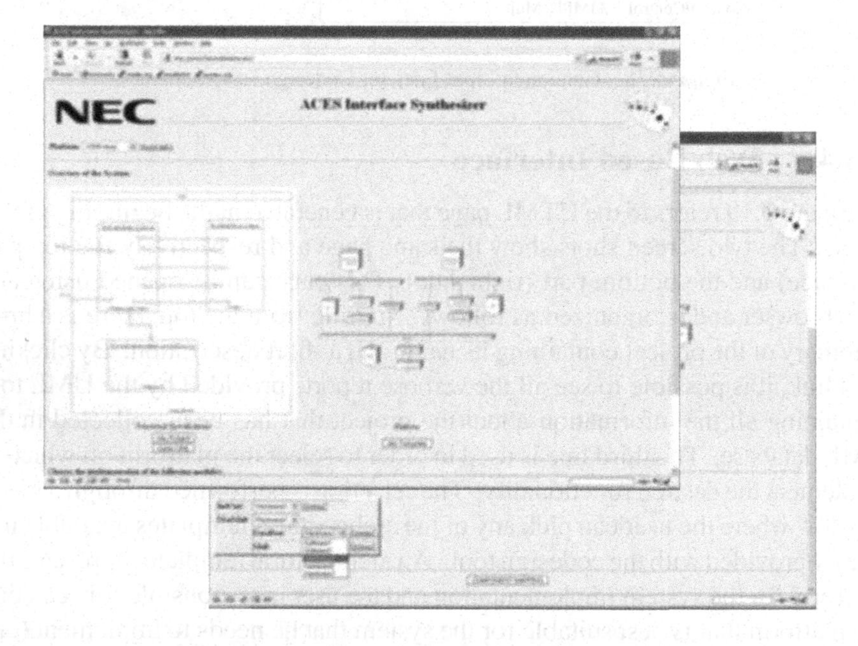

Figure 6.20. Mapping on a Dual Processor Architecture

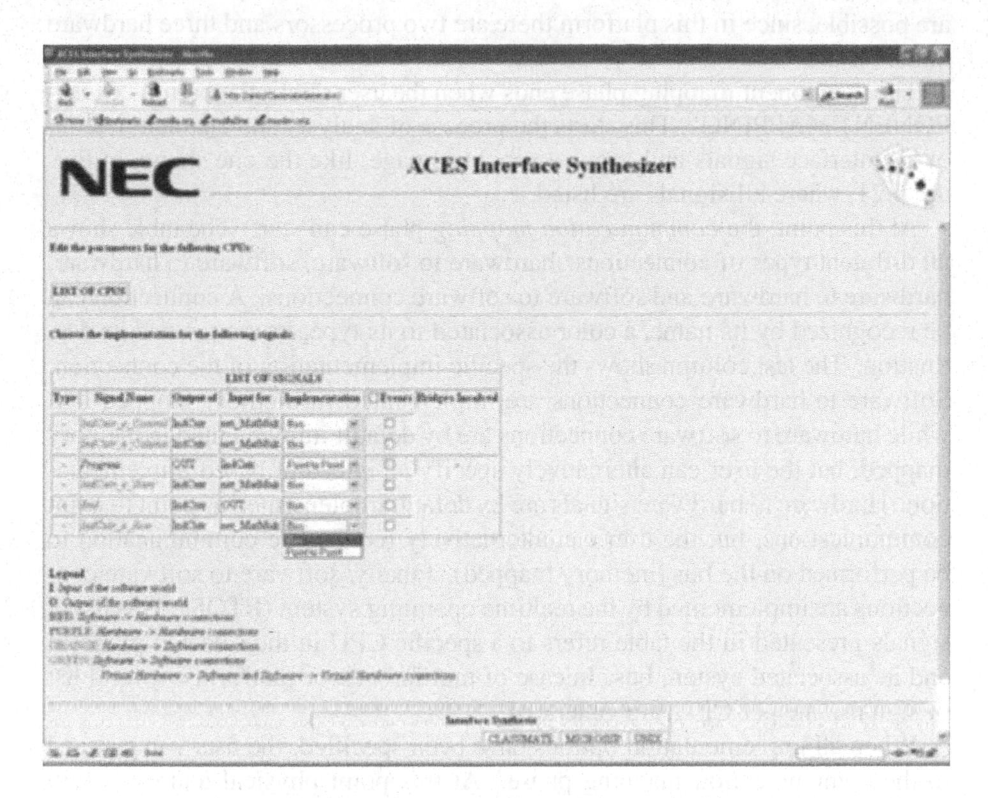

Figure 6.21. Communication Refinement

view of the system exported from the UML specifications and on the right side there is the picture of the selected platform. By changing the target platform, the picture on the right is automatically updated. For example in Figure 6.19 the platform contains only one processor, while in Figure 6.20 the platform contains two processors and a two level bus hierarchy. The idea behind this solution is to support a function architecture codesign approach that requires the separation of the functionality from the architecture selected for its implementation.

Finally, at the bottom of the page are listed all the objects present in the functional specifications and the user can specify the implementation (i.e., the hardware or software component of the platform onto which the functionality will be implemented.) for each of them through a menu window associated to each object. This is what we call *component mapping phase*. The number of choices available for this mapping depends on the selected platform. For example, in the platform shown in Figure 6.19, only two choices are possible (Software1, Hardware1) due to the fact that it is a simple single processor architecture with one hardware component connected to the processor bus. But in the multiprocessor architecture shown in Figure 6.20, five different choices are possible, since in this platform there are two processors and three hardware units.

The component mapping phase ends when the user clicks the button "COMPONENT MAPPING". This starts the process of analysis and characterization of all interface signals and opens a new html page, like the one shown in Figure 6.21, where all signals are listed.

At this point, the *communication mapping* phase can start. The table shows all different types of connections: hardware to software, software to hardware, hardware to hardware and software to software connections. A connection can be recognized by its name, a color associated to its type, its source and its destination. The last column shows the specific implementation of the connection. Software to hardware connections are implemented in memory mapped I/O, while hardware to software connections are by default implemented in memory mapped, but the user can alternatively specify an interrupt based implementation. Hardware to hardware signals are by default implemented as point to point communications, but the user can alternatively require the communication to be performed on the bus (memory mapped). Finally, software to software connections are implemented by the realtime operating system (RTOS). This list of signals presented in the table refers to a specific CPU in the selected platform and its associated system bus. In case of multiprocessor platforms, several list of signals, one per CPU, are generated.

When all implementation options have been specified, the user can proceed to the communication mapping phase. At this point physical addresses will be generated for all memory mapped communications and specific interrupt lines of the processor will be selected for signals implemented in interrupt. The result is a new page, not shown here, similar to Figure 6.21, but where the last column shows now the physical addresses and the interrupt lines that have been selected. After having examined all the communications, the user can still go back and change some implementation options or, if satisfied, proceed to the next hardware/software cosimulation phase.

6.7 Conclusions

The complexity of current embedded systems requires large teams of designers that interact especially at the early design stages when architecture selection and hardware/software partitioning take place. Models and tools that allow one to visualize and document the design abstractions and the interactions between different components or levels of abstraction of a specification are essential. UML being platform independent and with a rich graphical notation can serve this purpose. We presented a methodology that specializes the UML standard notation for modeling embedded systems platforms and protocols leading to an integration with an existing hardware/software codesign technology.

Chapter 7

UML Tailoring for SystemC and ISA Modelling

Giovanni Agosta, Francesco Bruschi, Donatella Sciuto

Politecnico di Milano
Milano, Italy

Abstract In this chapter, two approaches to the use of UML as a formalism for the design of digital systems are presented, based on the two metrics defined for the classification of design formalisms, abstraction and application specificity. According to the language features represented by these figures, we address two different modeling domains, defining specific profiles characterized by a low or high application specificity.

7.1 Introduction

Object oriented formalisms, after having been widely accepted in the software world, are making their way into the specification and design of complex hardware/software systems. The constantly growing complexity of these devices, in fact, adds new requirements to the modeling tools involved in the design flow. The Unified Modeling Language (UML) is a visual formalism for the design of object oriented systems that is gaining consensus as a result of its standardization and expressive versatility.

In this chapter we explore the possibility of exploiting UML in the design flow of hardware and hardware/software digital devices. The aim of the work is twofold:

1. to analyze the roles that UML can play with respect to other system formalisms;

2. to explore the possibility of effectively exploiting UML in the roles identified in the previous analysis.

To better understand and frame the benefits UML can provide to a SoC design flow, we rely on two evaluation metrics, *abstraction* and *application specificity*.

G. Martin and W. Müller (eds.), UML for SOC Design, 147-173.

These two concepts, even if not completely unrelated and orthogonal, allow one to clearly identify and highlight the peculiarities of a flexible high level language such as UML in the context of a hardware/software design flow. Moreover, the two different approaches presented for the exploitation of UML in a SoC design scenario will be characterized in terms of the two attributes introduced, in order to clearly understand their contribution to each one of them.

7.2 Abstraction and Application Specificity

A typical modern design flow can be seen as a series of steps, applied in a defined sequence, that aim at the implementation of a specific functionality. What typically happens is that some *desired behavior* must, at the end of all the design phases, be described in terms of elementary *manufacturing actions* dependent of the implementation technology chosen. The intermediate steps through which the designers can be required to go can be extremely heterogeneous from many different points of view. Nevertheless, it can be worth seeking some core concepts whose meaning remains defined throughout all the different stages, and that could help in defining some basic invariant properties of the design flow.

An alternative way of describing a design flow is to depict it as the production of a series of descriptions of the system, incrementally richer in information related to the implementation. Ideally the design would start from the pure statement of the functionality of the system to be realized, without any information on how this will be implemented. Then, as far as implementation choices are performed, new descriptions are produced which contain details that reflect such choices. Even if apparently straightforward, it is worth noting that an implicit constraint strictly imposed on the subsequent descriptions is that the functionality of each of those must remain the same.

This simple analysis implies a set of assumptions that are usually implicit, and that are worth being highlighted:

- the outcome of each design flow stage is some *description* of the system;
- descriptions of the same system at different stages differ in that they reflect a different amount of implementation details;
- all the descriptions of the same system are assumed to *describe* the same *functionality*.

These assumptions can be further made explicit as follows:

- it is assumed that it is possible to associate a *functionality* with every *description*. Note that this association may not be easy to determine;
- it is assumed that functionalities of descriptions at different stages can be compared for equivalence. This is a very strong assertion, since defining

an equivalence between heterogeneous models can be extremely difficult, but nevertheless it cannot be disregarded, since giving it up would conceptually prevent the possibility of checking for design *correctness*.

Note that two classes of concepts are present at every stage of a generic design flow: *descriptions* and *functionalities*. Examples of what is meant by *description* are:

- A textual description of the expected behavior of a system;
- A C function that maps an array of floating point numbers onto another;
- A netlist of elementary elements such as logic gates and flip–flops.

Examples of functionalities, on the other hand, are:

- a representation of the input/output relation of a low pass filter;
- the behavior of some observable feature of a system as a response to some input.

In the context of a design flow, what is usually meant by the term *model* is a *description* from which a certain *functional interpretation* is derived. Note that the nature of the functionalities considered depends mainly on the type of system or subsystem under analysis.

These considerations can be formalized as a mathematical relation between *descriptions* and *functionalities*.
A *model* is a pair $\langle d, fi \rangle$, where:

- d is a *description*, $d \in D$, and D is the set of all possible descriptions;
- $fi : D \rightarrow F$ is the *functional interpretation*, and F is called the *functionality space*. fi is injective (a description can have only one functionality, given a functional interpretation).
- $fi(d)$ is the *functionality* of D.

Since different functionalities can be equivalent with respect to some equivalence relation $e \in F^2$, for simplicity we will consider, as a functionality space, the space of all the equivalence classes induced by e. Since fi is injective, an equivalence relationship $e_D \in D^2$ is induced on D: $(d_1, d_2) \in e_D \Leftrightarrow f(d_1) = f(d_2)$. Amongst all the possible descriptions that form the domain of a given functionality space, several clusterings can be made. In particular, it is possible to group all the descriptions produced with a given *formalism* L (a software example could be: all the C programs; all the assembly descriptions; all the models of a processor given its memory content and state registers). The point is that some interesting properties of the different formalisms can be stated in terms of properties of the description subspaces and of the functional interpretation fi.

Let us define another function of a description $d \in D$:

$$I : D \rightarrow R^+$$

that represents the *information content* of the description d. The expression *information content* must be used according to Shannon's definition.

One of the most important features which a formalism for the system level design must have is the ability to express specifications that can be *easily* interpreted and analyzed in the early phases of a project by various different designers and analysts. This is a key point in enabling important possibilities such as model exchange, verification of correctness, and, most of all, communication between system level engineers and designers. The ease of interpretation of a description can be put into correspondence with the ability to evaluate its *functionality* by a human user. Greatly simplifying the complex compound of perceptive and psychological phenomena that lie behind the ability to interpret a description which extracts certain functional features from it, it is reasonable to state that the smaller is the description *information content*, the easier it will be to interpret it. It is then possible to compare the understandability of two descriptions d_1 and d_2 that *belong* to different formalisms L_1 and L_2, that have the same functional interpretation, by comparing their information contents: d_1 is *more easily understandable* than d_2, given that they have the same functionality $fi(d_1) = fi(d_2)$, if $I(d_1) < I(d_2)$. If this property is reflected by a consistent number of functionalities interesting for a given design domain, then, *in that domain*, L_1 is more easily *understandable* than L_2.

Another interesting feature of a *formalism* is its ability to represent functionalities easily. The *specifications representation problem* can be stated as follows. Given a functionality $f \in F$, where F is a given functional domain, and a formalism L, what is the effort in finding a description $d \in L$ such that $fi(d) = f$? Again, it is possible to formalize conceptually this problem by assuming that if in L_1 there is a description d_1 with a smaller information content than a description $d_2 \in L_2$, then it will be easier for a designer to find d_1 than d_2, or, in other words, to model f in L_1 than in L_2. A way of reinforcing this assumption is to remember that the *information content* of a description is, according to Shannon's interpretation, the number of modeling choices which must be performed to obtain it. If

$$I(d_1) > I(d_2), d_1 \in L_1, d_2 \in L_2, fi(d_1) = fi(d_2)$$

for a consistent number of functionalities of interest F, then L_1 is said to be *more expressive* with respect to L_2. Note that both expressiveness and understandability depend on the set of functionalities F considered.

Amongst all others, two features directly influencing *expressiveness* and *understandability* of a given *formalism* can be defined: its *abstraction* and its *application specificity*.

Abstraction is related to the amount of detail that must be provided, in a given formalism, in order to describe a given functionality. A way of defining abstraction differentially amongst different formalisms is the following: given a functionality $f \in F$, if there are more descriptions $d_{1i} \in L_1$ than $d_{2j} \in L_2$ such that $i(d_{1i}) = i(d_{2j}) = f$, then L_2 is *more abstract* than L_1 with respect to f. This definition is directly based on the etymological meaning of abstraction, in that it expresses the possibility, for a more abstract language, of describing a property common to several different descriptions in a less abstract language (in this case, the property is the functionality).

Application specificity is the possibility, for a given formalism, of effectively describing *functionalities* which belong to a specific application domain. This happens when a certain amount of information on the specific domain is *embedded* in the language definition. Descriptions d in a formalism that is *application specific* with respect to a subset $F_a \subset F$ will have a lower information content with respect to descriptions in a non–application specific formalism.

Both *abstraction* and *application specificity* are features that increase the expressive efficiency of a given formalism, that is, they allow the description of functionalities of interest with less information. Nevertheless, a high application specificity narrows the expressive domain of the formalism. Thus whilst abstraction has no drawbacks when present in formalisms adopted in the early phases of system level design, the application specificity can keep a formalism from being adopted in a wide range of applications.

The interesting point is that UML, whilst being abstract, can allow different levels of application specificity. In particular, its profiling features allow a specialization of the syntactical and semantical elements, importing concepts that are typical of a given set of applications.

In Section 7.3 an approach to the use of UML with a high degree of abstraction and low application specificity is shown. The specialization features of UML are employed to define a set of concepts present in the *Transaction Level* communication modeling style. Transaction Level Modeling allows the description of communication between modules of a system disregarding information that is implementation related, such as the protocol and the semantics of the communication means. The semantics are similar to that of the *remote procedure call* (RPC). The information content $I(d)$ of a Transaction Level description d is typically much lower than that of a description of a system with the same communication functionality written, for instance, in VHDL, where explicit synchronization and acknowledge information must be specified. On the other hand, there is no specific application domain concept in Transaction Level Modeling, that can be applied to describing a wide range of systems.

Native unconstrained UML model elements are more abstract than the core concepts of Transaction Level Modeling. Section 7.3 shows how to constrain UML elements in order to "mimic" the concepts typical of an abstract textual formalism for the specification of communication between functional elements. The result is the formalism L_{TLM}.

On the other hand, in Section 7.4 the problem of modeling a narrow set of systems is analyzed, in order to verify the possibility of effectively using UML while varying application specificity. The set $F_{ISA} \subseteq F$ considered is that of instruction level programmable systems. A set of concepts functional for the description of these components is defined by specializing core UML elements, and these are applied to the description of existing instruction set architectures. The set of these concepts, together with the syntactical rules for their composition defines the formalism L_{ISA}. It is interesting to compare the expressive effectiveness of the two profiles: the information content of a description d whose functionality lies in F_{ISA} would be much higher if $d \in L_{TLM}$ than if $d \in L_{ISA}$. On the other hand, there is a wide range of functionalities f for which a description d does not exist in L_{ISA}.

The comparison of pros and cons of both approaches will allow an evaluation of the effectiveness of UML when used considering different levels of application specificity.

7.3 UML Transaction Level Modeling

In this section we present a profile that defines within UML a set of elements typical of Transaction Level Modeling. Through the UML specialization mechanisms we formalize the concepts of module, channel, and event based synchronization. In addition to the possibility of modeling the communication structure of a system, we consider the possibility of modeling behavioral aspects by means of state diagrams as part of the UML formalism. This is a substantial extension of the work in [27].

Having defined these elements that allow the composition of an executable model of the system to be designed, we face the problem of automatically generating code from the model. The problem is tackled at both the conceptual (mapping from the UML model semantics to design language semantics) and technological level (choice of portable and standard technologies).

The design language chosen as target for the translation is SystemC[161], and the translation flow is fully based on standard technologies such as XMI [145], XSLT[225], DOM[48], SAX[183].

This section shows how UML can be employed at a high abstraction level, low specificity level design formalism by means of a case study in which, starting from a graphical model, a SystemC description is generated.

An interesting question is whether the modeling capabilities of UML can be applied to embedded system design, and integrated in a flow that comprises SystemC 2.0 [161] as the modeling language. In particular, such a flow should allow the use of the high level modeling features of UML in the early phases of the design, and then it should be able to map this information onto a SystemC model. This approach can provide several advantages. Most of them are well proven in the software design field:

- Using a visual design approach lets the designer focus on the essential architectural and functional features of the system in the early phases of the project, without being bothered with the many details (syntactical, for instance) of a textual design language, that is to say, that the abstraction is higher;

- Visual models are a documentation mean of proved effectiveness; the project documentation task can thus be simplified by the adoption of a visual modeling approach;

- A point of great importance is the possibility, given by a modeling language such as UML, of locating architectural patterns and expressing them for further reuse. The pattern idea extends the reusability concept from the object to the architecture domain, and is becoming widely used in the design of complex software systems. A great advantage coming from the integration of UML in the design of hw/sw systems would be to explore whether such concepts are meaningful in the context of the design of embedded systems; this would be of great interest for the management of the ever growing complexity of the design of this kind of systems.

We divide the problem of representing a system design in UML in its structural and behavioral components. In Section 7.3.1 we define a profile for the structural description of *Transaction Level* models, whilst in Section 7.3.2 we augment the system description with behavioral elements that describe the functionality of each component of the system.

The expression *Transaction Level Modeling* (TLM) refers to an abstraction level in the description of a system which provides modeling of the communication between the elements that describe the behavior of the system in a functionally, but not *pin-accurate* way. That is, in a TLM model the focus is on the data which is passed between two modules, rather than on the way the transfer is accomplished.

For instance, it is possible to specify the functional characteristics of the communication, such as the blocking or non–blocking semantics, without defining their implementation. To do so the designer does not need to use the hardware signal semantics, as happens in languages such as SystemC 1.0 and VHDL.

One of the many advantages of the introduction of such a modeling style in a specification language is the possibility of obtaining an executable model at a higher level of abstraction, not biased by architectural choices. Most of the implementation choices will be performed after this early modeling phase. Thus, a TLM model of a system can describe an abstract system that can be mapped onto different architectures.

Transaction Level Modeling was first introduced in hardware specification languages in SpecC [235], and later developed under the name of behavioral wrappers in [234] and as Functional Interface by the VSIA [118].

7.3.1 Structural Features of Transaction Level Models

The first step in defining a UML profile and toolchain to describe HW/SW systems is to define their structural components. We first define a UML profile, then we describe the flow of code generation.

Profile Definition. The profile has been defined to satisfy the characteristics previously required, by exploiting the *stereotype* extension mechanism of UML. A stereotype is used to tailor UML constructs to the needs of specific application domains.

In Figure 7.1 the elements of the profile and the relations between them are shown as a UML class diagram.

The stereotypes defined correspond to the conceptual entities used to capture the communication features offered by SystemC:

- The `<<module>>` stereotype is intended, in the profile, as the basic encapsulation element; it acts essentially as a container of processes and of other modules. Moreover, the possible communication links amongst modules are different from those among processes. In this way the encapsulation features typical of the SystemC 2.0 modules are preserved; the modules can act as sender and receiver of messages, and can communicate with other modules by means of `<<moduleLink>>` associations.

- The `<<process>>` stereotype represents the behavioral elements of modules. Two processes can communicate directly only if they belong to the same module; communication between two processes of different modules is achieved by means of intermodule communication links. The processes can act as sender and receiver of messages, which in turn are realized by `<<processLink>>` associations. The `<<process>>` stereotype is the top of a hierarchy that comprises elements corresponding to the SC_METHOD, SC_THREAD, SC_CTHREAD SystemC process types.

- The `<<message>>` stereotype is used to represent information exchange between different modules and processes. These are the direct links to the

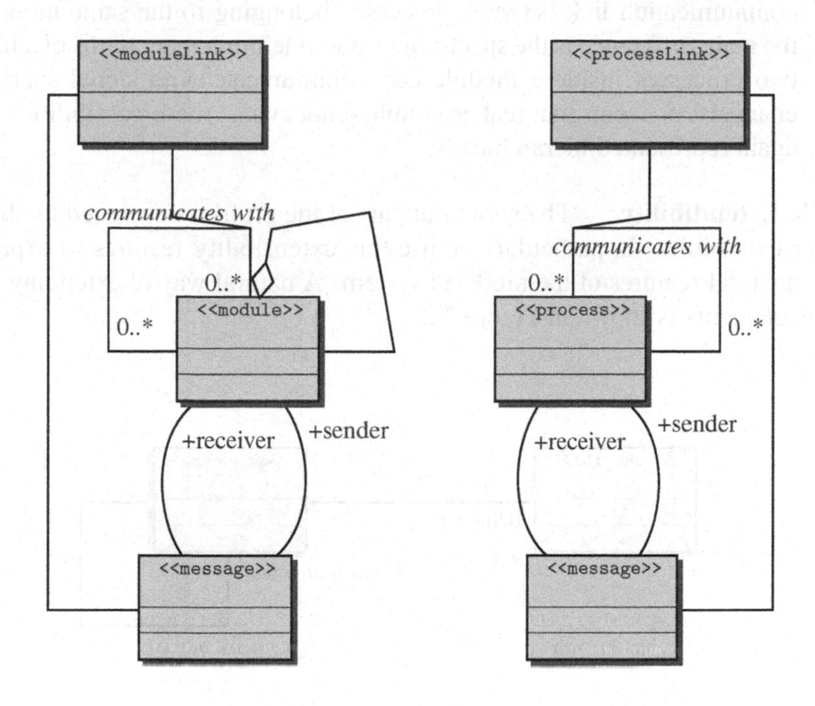

Figure 7.1. Relations between Profile Elements

collaboration diagrams obtained from the UML design phases: for every message between two entities in the collaboration or sequence diagrams, there has to be a corresponding `<<message>>` in the class diagram. Messages must be associated with `<<moduleLink>>` or `<<processLink>>` classes, according to the nature of their senders and receivers (either modules or processes). This association is the link between the UML collaboration diagrams and their SystemC realization.

- `<<moduleLink>>` stereotype represents the "links" that implement the exchange of a set of messages between two modules. In SystemC this concept corresponds to that of **channel**. In SystemC the channel entity is specialized into less general, lower level specializations: the `<<module Link>>` has the same characteristic. This isomorphism is meant to give control over the code generation phase: the designer can decide to use a signal to realize a set of messages instead of a more general channel; this information will be reflected in the generated code.

- <<processLink>> is analogous to <<moduleLink>>: it represents a communication link between processes belonging to the same module; the main difference is the spectrum of possible implementations of a link: two processes inside a module can communicate with signal sharing, channels, or events that realize simple rendezvous; these possibilities are again represented hierarchically.

Profile Extendibility. The structural part of the profile is designed to allow further extensions. In particular, we use this extensibility features to express the behavioral features of the modeled system. A natural way of extending the profile elements is shown in Figure 7.2.

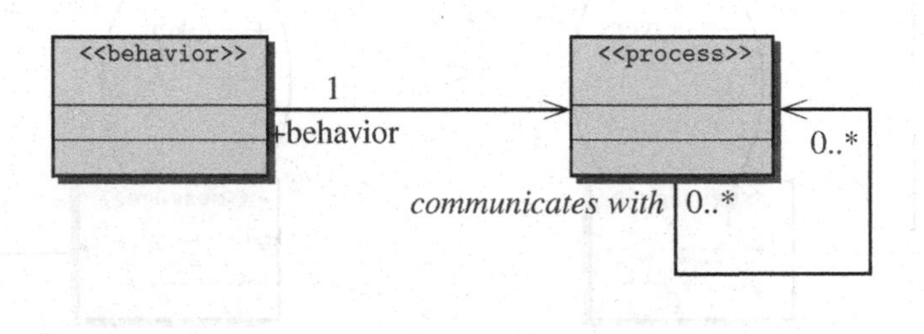

Figure 7.2. Process Behavioral Extension

Here the <<process>> class is associated with a <<behavior>> class, which in turn can express some behavioral properties of the process (for instance, it could be associated with a State Machine diagram).

The link stereotyped classes (<<processLink>> and <<moduleLink>> classes) are susceptible to a similar extendibility: there could be, for instance, a set of communication protocols that can be attached to a channel and then synthesized in the code generation phase. The profile elements and associations are defined in order to allow such extensions.

Code Generation Flow. The proposed design process comprises a UML design phase, a refinement phase that extracts from the UML model the information needed for code generation using the concepts defined in the profile, and two automatic translation phases, that operate a series of transformations to obtain the final code. The implementation of the flow implies the use of different emerging technologies in the field of data exchange:

UML model (collaboration, sequence, class) → UML profiled class diagram. This is the translation phase in which the designer, after having outlined a suitable set of communication scenarios, distills the information needed and expresses it in terms of the concepts defined by the profile. The steps needed to perform the translation are:

- identify the module-process architecture (i.e., assign every process to a module);
- for each link between two processes in a collaboration diagram:
  1. define a set of `<<processLink>>` association classes if the processes belong to the same modules, a set of `<<moduleLink>>` between the containing modules otherwise;
  2. assign each `<<message>>` that connects two processes to a link;
- repeat for each module.

UML profiled class diagram → XMI model description. This step is performed by the UML modeling tool; XMI (see [145]) is a XML format that is standardized by OMG (see [144]); it allows the exchange of design models. XMI provides data exchange not only among UML modeling tools: it has the capability to represent every design model whose metamodel is described in terms of the OMG Meta Object Facility (MOF) (see [149]). Most of the UML tools now available include an XMI generation module that allows the export of the model in compliance with this XML format;

XMI model description → XML intermediate format. The XMI representation of a UML model is very rich in details that relate to things such as the graphical representation of the elements, the references between objects in different diagrams, and so on. Moreover, the data are generated according to the MOF metamodel structure of the UML language: this means that the information associated with the profile elements is not easily accessible. Therefore this format is not an ideal starting point for the code generation; so a choice was made to perform a first transformation on the XMI representation, in order to extract from it only the relevant details needed by the next phases. Another significant choice was to obtain from this transformation another XML compliant document. This, in fact, allows an easy data parsing by the subsequent algorithms and a much easier data exchange with third parties' tools. The technology chosen to perform this step is the W3C XSLT (see [225]). XSLT is a set of recommendations for a scripting language that is able to transform a XML document into another XML document by means of a sequence of transformations. The XLST scripts are XML documents themselves. This translation phase is then accomplished by means of an XLST script, whose main tasks are:

- to extract the model information needed to build the intermediate format;
- to format the information retrieved in a useful fashion.

To achieve the first goal the algorithm has to retrieve all the instances of the stereotyped classes, the associations between them, and to output all the related information, formatted as an XML document. This intermediate format contains a list of modules, each one in turn containing a list of processes; for each process there is a set of references to each process which exchanges messages with it, together with the message signatures and the links to what they belong.

XMI intermediate model → SystemC skeleton code. This step can be again performed using XSLT transformations; the intermediate description can also be easily parsed using an XML parser and then elaborated in order, for instance, to compute some metrics from the static information contained in it.

7.3.2 Behavioral Description of Transaction Level Models with UML State Diagrams

In this section we define a behavioral extension of the structural profile previously defined, based on the State Machine UML diagrams.

As stated in Section 7.3.1, the description of the structural elements of a model can be augmented with arbitrary information by means of the extension syntax proposed. We exploit this possibility in order to associate behavioral functionality descriptions with modules. In particular, we have chosen the State machine diagrams defined in UML 2.0 as the computational model of the modules' behavior. This choice is somewhat arbitrary, since, in principle, other computational models could be adopted for the same purpose. Nevertheless, State machines are directly available in UML and their expressiveness is adequate for a significant set of application domains. When a process behavior is to be specified the behavioral extensibility is exploited by associating a State Machine with the process. Interaction with other processes and with external modules is represented by the *transitions' triggering*: the set of all possible triggers is naturally associated with a State machine. In the SystemC implementation these triggers are implemented either as `<<moduleLinks>>` or as `<<processLinks>>`. In the first case it must be possible to fire the triggers from outside the module. Thus the triggers must be accessible as *interface methods*. In the second case, triggers are implemented by a couple of `<<processLink>>`: the notification of an event and the modification of a variable visible at module level.

The main issue in extending the model with State machines behavioral information is to define a proper translation of the UML diagram semantics to the target language, in this case SystemC 2.x. There are different possible implementations of the semantics considered with the behavioral concepts of

SystemC. Out of all the possibilities, the following translation rules were chosen:

- for each *state* in a *State Machine* associated with a process, a SystemC *thread* is generated. The reason why states are represented by threads is to allow parallel state activation semantics, present in State Machines;

- the activation of each state is represented by a boolean signal. More than one state can be active at the same time;

- for each possible trigger an event is instantiated. All the state threads are sensitive to the notification of every trigger event that can possibly fire a transition from that state;

- a variable `last_trigger` that identifies the last trigger fired as an enumerated value is instantiated;

- a trigger fire is implemented as an event notification and as a change of the `last_trigger` variable;

- when a trigger is fired all the states which are sensitive to it are awakened; if they are active the triggered transitions are executed if the corresponding guarding conditions are true.

As an example consider the fraction of a State machine shown in Figure 7.3. The code structure of the thread implementing state 1 is shown in Figure 7.4.

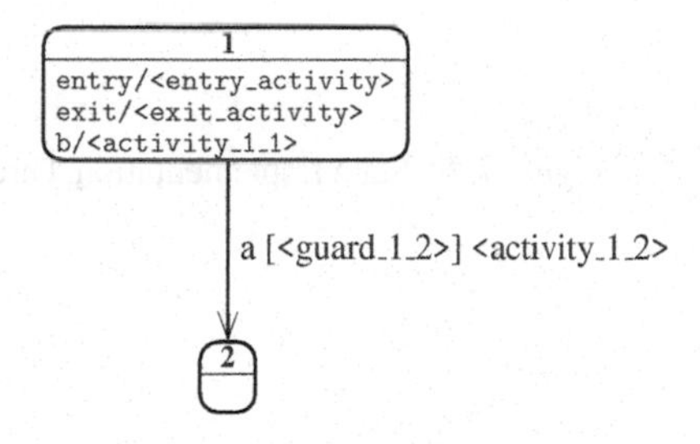

Figure 7.3. State Transition Instance

```
void state_1_thread() {
    while(true) {
        wait();
        if (state_1_active) {
            state_1_entry_action();
            switch (current_event) {
                case event_a_h:
                    if (guard_1_2) {
                        activity_1_2();
                        state_1_active=false;
                        state_2_active=true;
                        break;
                    }
                case event_b_h:
                        activity_1_1();
                        break;
            }
            state_1_exit_action()
            current_event=no_event;
        }
    }
}
```

Figure 7.4. State Implementation Thread

7.4 Application Specific UML Modeling

In this section we explore the possibility of using the specialization mechanisms of UML to define conceptual toolsets that specifically target an application field, such as multimedia processing or processor design.

First, we evaluate the potential effectiveness of this approach from a theoretical point of view. Then, we support our analysis by means of a case study targeting the field of instruction set architecture design.

In the case study, UML is used to describe the typical concepts used in the definition of instruction set architectures. A profile is defined, and its use is exemplified by modeling some sample instruction sets.

7.4.1 Motivation for Highly Specific System Design

As we have seen in Section 7.2, high specificity is one of the characteristics which allows a design to be easily understood by an observer. When the observer knows the application domain many application–specific details need not to be made explicit in the description, since the observer's knowledge will "fill the gaps" in the description.

However, a description cannot be simply underspecified, since this will make it understandable only to the observer that has application specific knowledge and abstraction abilities. Therefore what should be done is to create a specialized description language that has highly specific primitives, allowing a concise but well defined description of a system within a given application domain.

From the definition of Application Specificity given in Section 7.2 we now consider the mechanisms that UML 2.0 offers to customize the modeling language for the description of highly specific systems. The main mechanism offered by UML for specialization of a metamodel are the profiles.

By means of profiles we can describe highly specific aspects of an application, while preserving the high level of abstraction offered by UML.

7.4.2 Case Study: A UML Profile for the Description of Processor Instruction Set Architectures

To evaluate the effectiveness of UML profiles for the description of highly specific systems, we build a profile (the *ISA_profile* package) for processor *instruction set architectures* (ISA).

Instruction set description languages can be classified as structural and behavioral (see [167]). Behavioral languages abstract from the architecture, and directly describe the ISA semantics. This is the abstraction level at which *ISA_profile* works.

For the purpose of the *ISA_profile*, a processor ISA is divided into five components:

- data types;
- microinstructions;
- ISA syntactic specification;
- ISA semantic specification;
- register file and other implementation components.

Data Types. The basic elements of the description are the data items managed by the processor. In our description, these are always vectors of bits. Therefore we define a `<<BitVector>>` stereotype that becomes the root class for all data types used in the description of a processor ISA. The *Type* abstract class defines a bitvector object with basic operations.

We then define two levels of data type descriptions. First, there is a level at which the only relevant information is the information content of the data. This level is characterized by the stereotype `<<ISADataType>>`, which defines a size attribute.

Then, we add a level that takes into account the nature of the data – e.g., it allows the distinction of constant items, such as an immediate operand, from variable items such as registers. This level is characterized by the template classes *RO_object* and *RW_object*, stereotyped with `<<DataTempl>>`. The former defines data items with read primitives, whilst the latter inherits from *RO_object* and adds write methods.

Figure 7.5 shows the definitions of all the stereotypes required to define data items and types. In addition to the main items mentioned above there are a few more elements to consider:

- `<<DataItem>>` is the stereotype used to characterize data items;
- `<<ConcreteItem>>` is used to define an architectural component, such as a special purpose register, by means of the stereotyped generalization `<<ph_impl>>` from a data item;
- `<<composed>>` is used to stereotype associations of `<<ConcreteItem>>` objects — i.e., compound registers derived from the composition of shorter registers as in the "extended" registers of the Intel x86 family.

Microinstructions. To define the semantics of the ISA elements, we chose an operational specification. Therefore, the definition of the functionality of an instruction will be given as a State Machine whose actions are simple atomic operations, called *microinstructions*. This allows all defined ISA to be expressed in terms of the microinstruction language.

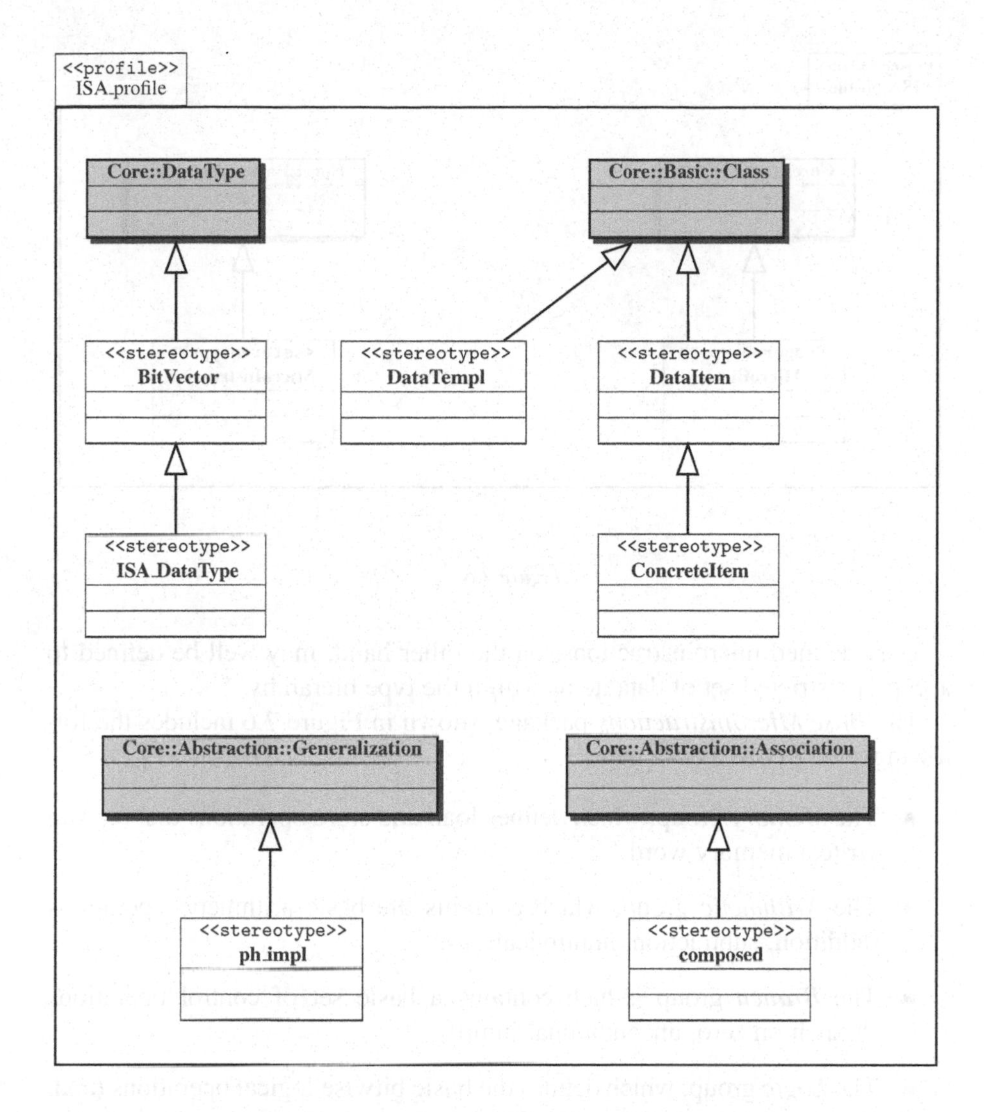

Figure 7.5. Stereotype Declarations of Elements Used in the Definition of Data Types and Items

Microinstructions are defined through a stereotyped class `<<MicroInstruction>>`, and can be collected into several broad categories defined by the `<<MicroBlock>>` stereotype.

The *Semantic_facilities.Base_MicroInstructions* package defines a set of basic microinstructions. Since the basic microinstructions are an abstract representation of functionality they do not work on actual data items. Rather they accept arguments of a single type, that is, instances of the *Type* class.

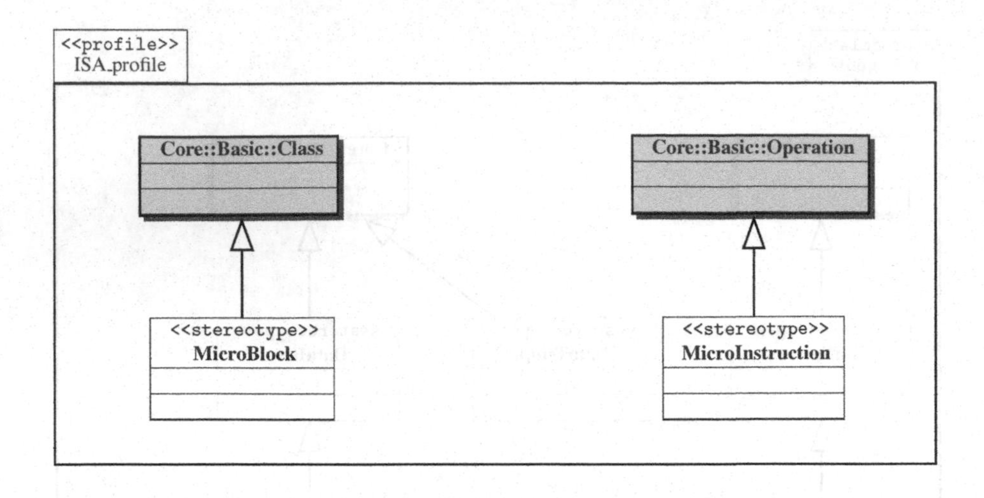

Figure 7.6.

User defined microinstructions, on the other hand, may well be defined to accept a restricted set of data items within the type hierarchy.

The *Base_MicroInstructions* package, shown in Figure 7.6 includes the following `<<MicroBlock>>` items:

- The *Memory* group, which defines load and store operations that read an write a memory word;
- The *Arithmetic* group, which contains the basic arithmetic operations (addition, subtraction, multiplication);
- The *Branch* group, which contains a basic set of control operations (branch on zero, unconditional jump);
- The *Logic* group, which defines the basic bitwise logical operations (and, or, not) and bit operations (shift).

ISA Syntax. The syntactic definition of the ISA provides the description of all the instructions formats allowed in the described processor. Instructions must be defined through the stereotype `<<Instruction>>`.

An instruction class is associated with the required operands (both sources and destination) via stereotyped associations. These allow the description of the following items:

- `<<source>>` describes a source register or immediate operand;

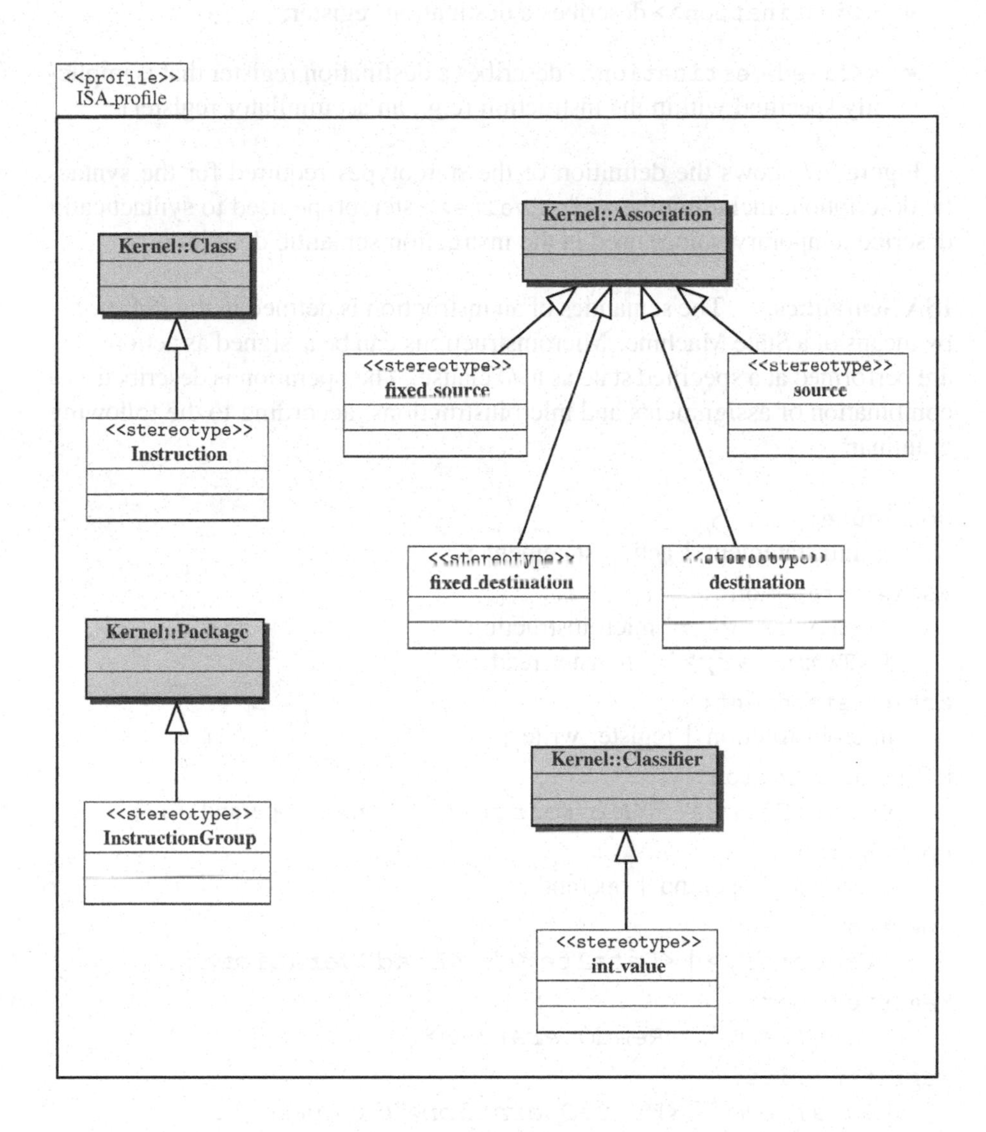

Figure 7.7. Specification of the Stereotypes Used in the Syntactical Definition of the ISA

- <<fixed_source>> describes a source register (or possibly immediate operand) implicitly specified within the instruction (e.g., an instruction which reads only from a specific register, as in a CISC processor);

- <<destination>> describes a destination register;

- <<fixed_destination>> describes a destination register that is implicitly specified within the instruction (e.g., an accumulator register).

Figure 7.7 shows the definition of the stereotypes required for the syntactic description, including the <<int_value>> stereotype used to syntactically describe temporary values used in the instruction semantic definition.

ISA Semantics. The semantics of an instruction is defined in the *ISA_profile* by means of a State Machine. Microinstructions can be assigned as actions that are performed at a specified state as a *do* clause. The operation is described as a combination of assignments and microinstructions, according to the following grammar:

```
do_clause:
    assign_statement | action_statement ;
assign_statement:
    <Temporary> '=' microinstruction
    | <Temporary> '=' register_read ;
action_statement:
    microinstruction | register_write ;
microinstruction:
    <MicroBlock> '.' <MicroInstruction> '(' operands ')' ;
operands:
    operands ',' operand | operand ;
operand:
    <Temporary> | <DataItem> '.' <ReadOperation> ;
register_read:
    <DataItem> '.' <ReadOperation> ;
register_write:
    <DataItem> '.' <WriteOperation> '(' operand ')' ;
```

Temporaries must be described in the syntactic specification, using the data types available in the design.

System Components. Some components of the system must be specified at least partially in order to allow the designer to define the ISA. For example, the register file should be known. This is required to allow the use of specific registers such as an accumulator.

Registers are defined as classes stereotyped with <<ConcreteItem>>, in order to distinguish them from the non–specialized data items. The <<Concrete Item>> stereotype points to the fact that the registers are elements of the structural description of the processor architecture rather than items of the conceptual description of the instruction set.

7.4.3 Modeling Examples of the Defined Profile

To prove the effectiveness of the defined profile we applied it to several architectures. We present here some significant parts of the MIPS [135] specifications.

MIPS Model. The MIPS is a RISC processor; it has a register file of 32 64 bit general purpose registers, used for both integer and floating point values. Memory addresses are 64 bit long.

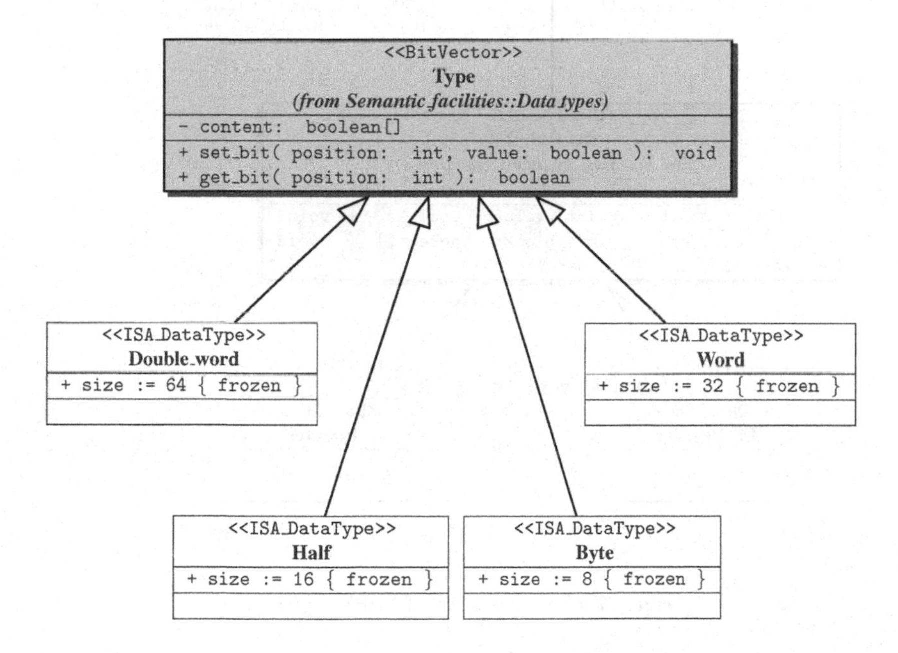

Figure 7.8. Specification of MIPS Data Types

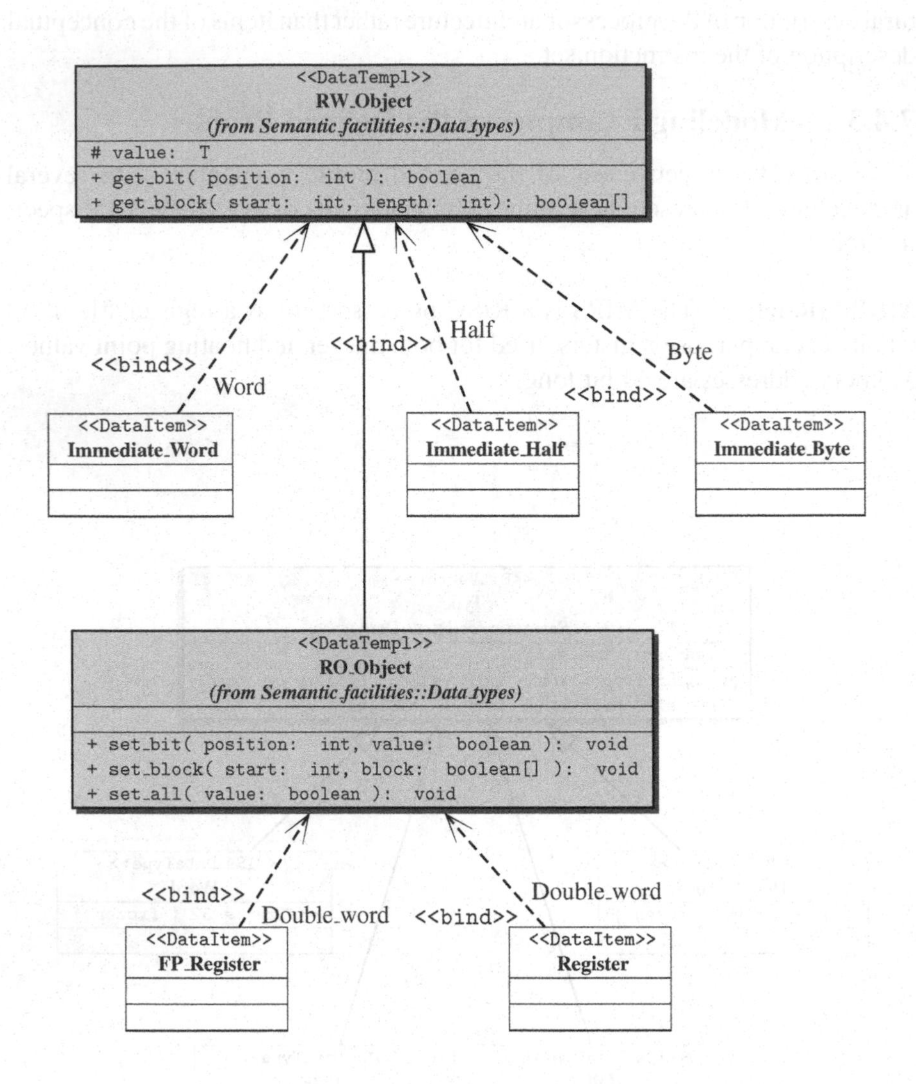

Figure 7.9. Specification of MIPS Data Items

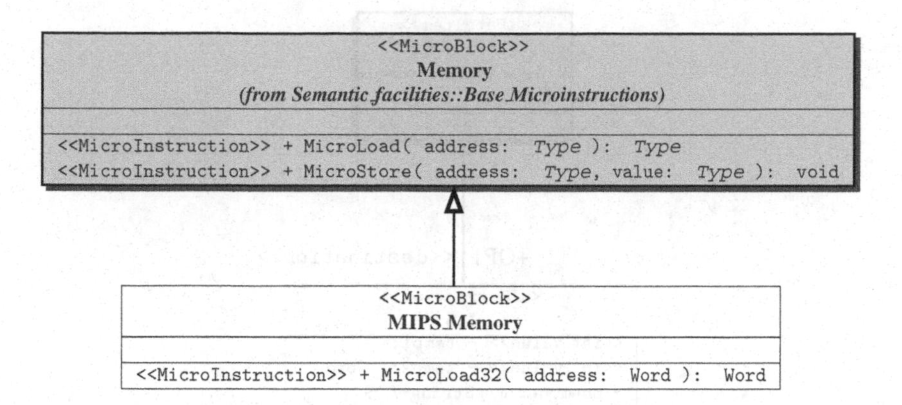

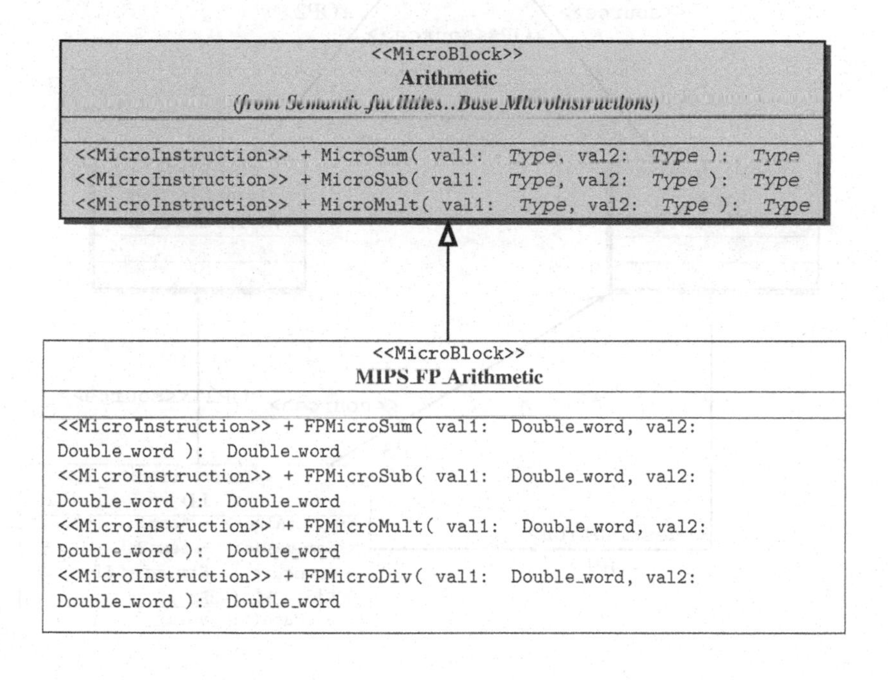

Figure 7.10. MIPS Semantics: Microinstruction Extensions

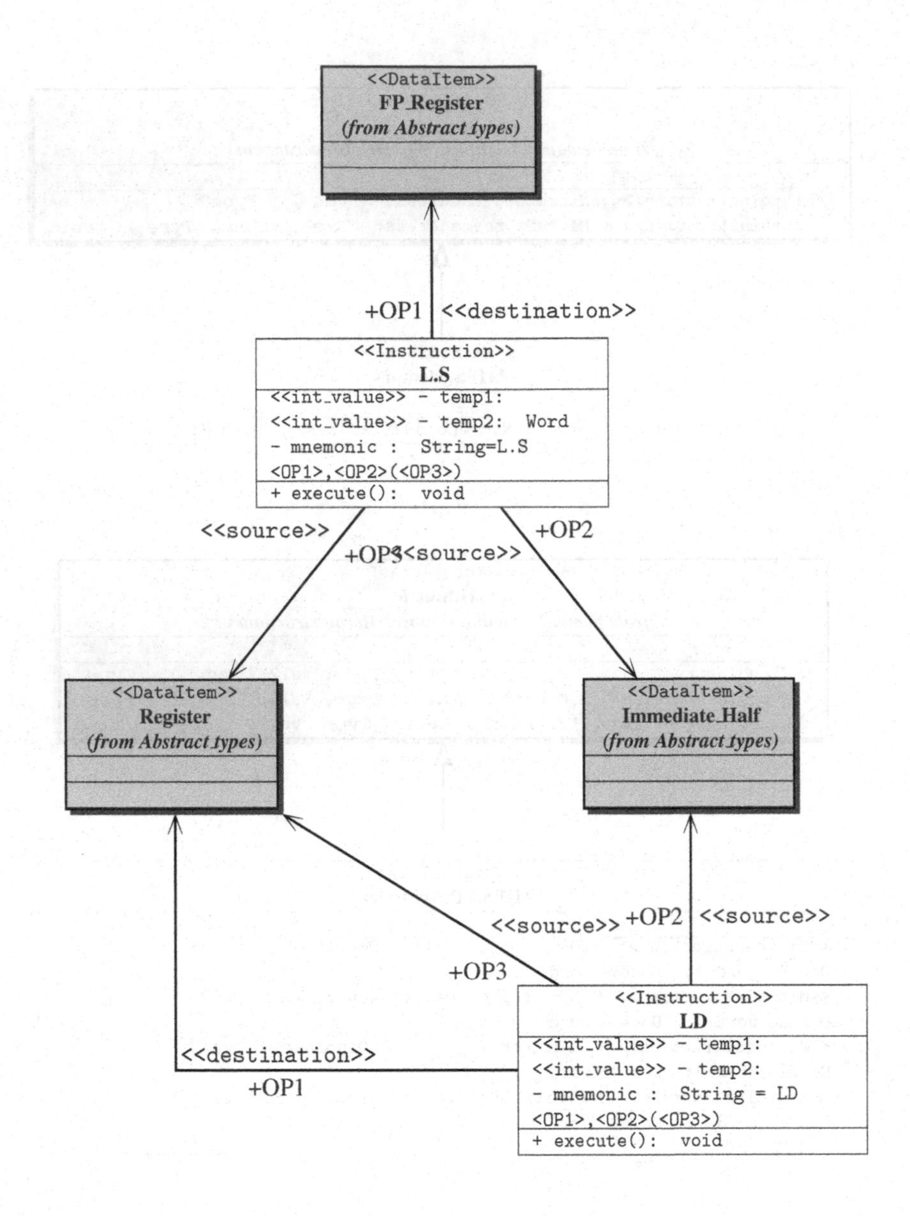

Figure 7.11. Example of Instruction Syntax for the MIPS

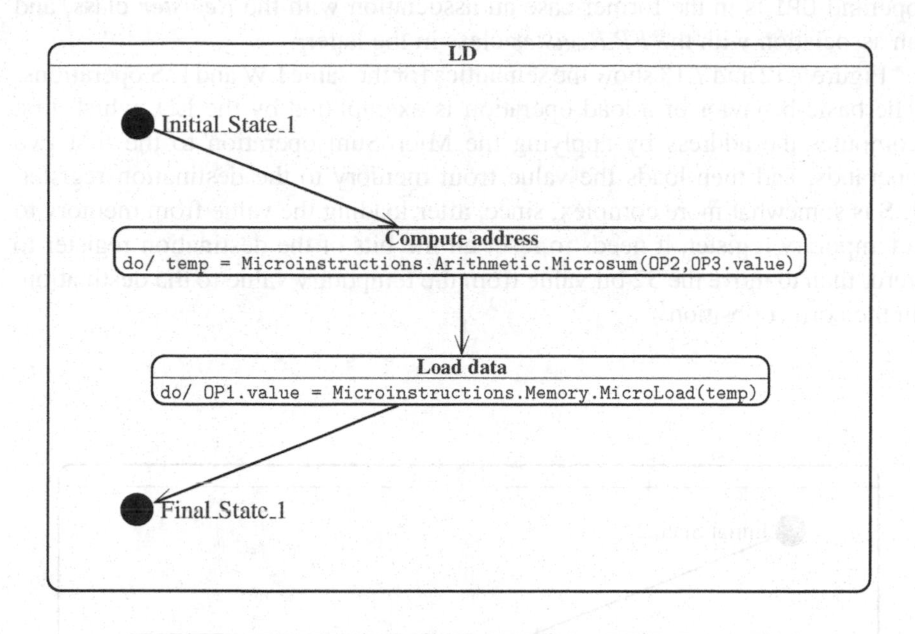

Figure 7.12. Example of Instruction Semantics for the MIPS: LD

Figure 7.8 shows the definition of new data types for the MIPS specification. The *Byte*, *Half*, *Word* and *Double_Word* types are declared as the basic units of information available to the machine. They are bound to the appropriate types through the `<<bind>>` stereotype.

The data items, on the other hand, are shown in Figure 7.9. Read only objects include the available immediate operands (from byte to word), while the read-write objects include two logical types of registers, integer and floating point. These are mapped, in the architecture's implementation, to the same physical register set, but are considered as separate in the abstract specification of data types.

Figure 7.10 describes the extensions to the set of microinstructions needed to specify at a high level the floating point operations. Other microinstructions are created to define operations that work on specific types (e.g., MicroLoad is specialized into MicroLoad32 and MicroLoad64).

Figure 7.11 shows an example of the instruction syntax for the MIPS, the definition of two load operations. Both operations have three operands: two sources and a destination. In both cases operands `OP2` and `OP3` are the source's immediate value and the register used in the address computation, while operand `OP1` is the destination register. While the LD operation loads a word as an integer, L.S loads a single precision (32-bits) floating point value. Therefore, operand OP1 is in the former case an association with the *Register* class, and an association with the *FP_Register* class in the latter.

Figure 7.12 and 7.13 show the semantics for the same LW and L.S operations. The basic behavior of a load operation is exemplified by the LD, which first computes the address by applying the MicroSum operation to the first two operands, and then loads the value from memory to the destination register. L.S is somewhat more complex, since, after loading the value from memory to a temporary register, it needs to reset all the bits of the destination register to zero, then to move the 32 bit value from the temporary value to the destination, in the correct position.

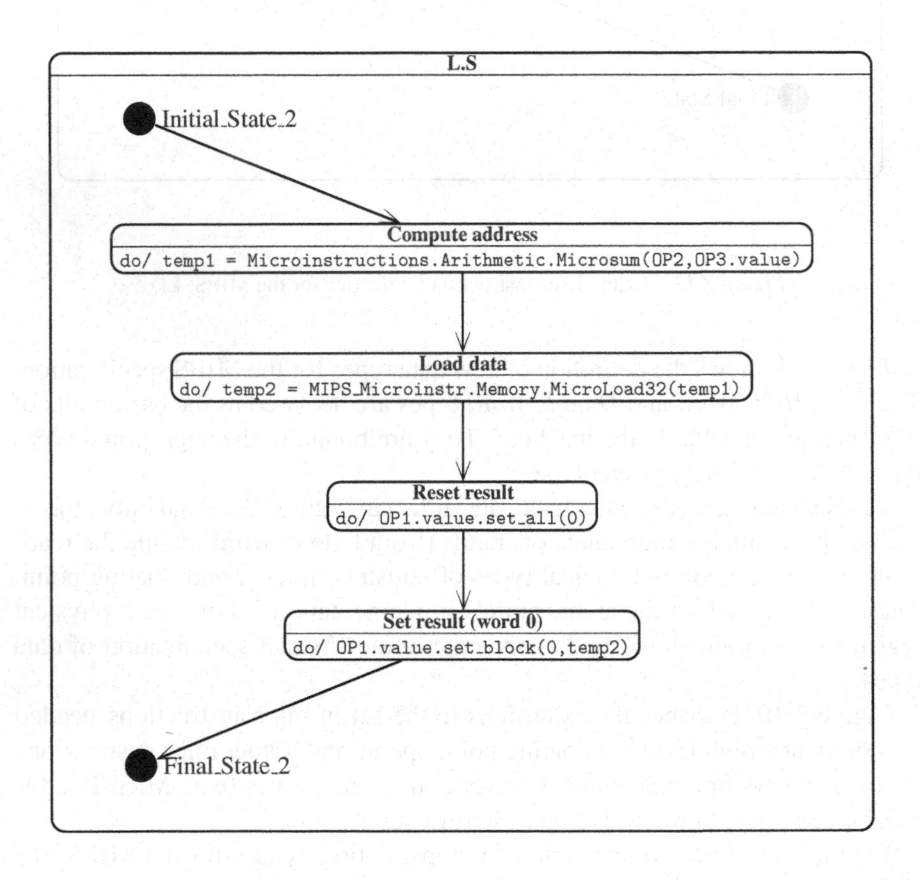

Figure 7.13. Example of Instruction Semantics for the MIPS: L.S

7.5 Concluding Remarks

The adoption of a high level formalism for the functional specification of systems appears to be effective even according to a formal analysis of the design languages based on the newly defined concepts of *abstraction* and *application specificity*.

As a low specific modeling domain we have chosen Transaction Level Modelling, which allows one to abstract, in the system level design, implementation details of communication between elements. We have enriched an existing approach with the possibility of specifying behavioral features by means of State Machines. We have then explored the possibility of translating the information present in the UML model into a SystemC 2.x description.

As a highly specific modeling domain we have chosen the description of processor instruction set architectures. We have defined a UML profile to capture the information related to the application domain, and have shown an application of the profile to the description of the MIPS ISA.

From the modelling experiments conducted UML proved to be effective in modeling at different levels of abstraction and application specificity.

Chapter 8

Model-Driven SoC Design: The UML-SystemC Bridge

Kathy Dang Nguyen, Zhenxin Sun, P.S. Thiagarajan, Weng-Fai Wong

School of Computing
National University of Singapore
Singapore, Republic of Singapore

Abstract We present a system level description mechanism based on UML notations from which one can automatically extract SystemC code. Our modelling framework is based on a restricted set of UML diagram types together with some extensions developed using stereotypes. As a result, applications as well as platform features can be captured at this level. Our system models are developed using the UML compatible tool, Rhapsody 4.2 [172].

8.1 Introduction

System level design methods for Systems-on-a-Chip seem inevitable given the technological trends and the accompanying economic pressures. In the recent past, a broad consensus has emerged regarding the basic principles that should govern system level design methods. Some of these principles are:

- The design methodology should support and deploy substantial component reuse.
- There should be an intermediate representation level with a clean *executable semantics*, at which both the application and the platform on which the application is to be realized can be captured and related.
- Behaviors described at the intermediate level, should clearly separate the computational aspects from the communication features.
- This intermediate representation should serve as a common design document for the software and hardware teams which can then independently work towards a detailed implementation.

G. Martin and W. Müller (eds.), UML for SOC Design, 175–197.

Given this wish list, two crucial choices to be made are the high level system description language and the intermediate representation language. We claim that a modeling language based on *UML (Unified Modeling Language)* notations for high level system descriptions and *SystemC* as the intermediate representation language constitute sound choices. Our main goal here is to substantiate this claim.

UML is now widely accepted in the software engineering community as a common notational framework. It supports object oriented designs which in turn encourage component reuse. It can be used to provide multiple views of the system under design with the help of a variety of structural and behavioral diagrams. It allows standard ways of extending the language to meet the demands of specific application domains. Though it was originally created to serve the software engineering community, UML is also becoming an attractive basis for developing system descriptions in the (real time) embedded systems domain [114]. In fact, many of the enhancements to the UML 2.0, the new standard, are geared towards easing the task of specifying complex real time embedded applications.

SystemC on the other hand, allows both applications and platforms to be expressed at fairly high levels of abstraction while enabling the linkage to hardware implementation and verification. Furthermore, SystemC — viewed as a programming language — is a collection of class libraries built on top of C++ and hence is naturally compatible with the object oriented paradigm that the UML is based on. Though SystemC is at present mainly oriented towards hardware descriptions, the enhanced version in the making [79] will support software module descriptions and run time features including scheduling. Hence SystemC has the potential to provide a full fledged description of an execution platform which can serve as the target of a codesign methodology. Thus SystemC is a viable intermediate representation language.

One might wish to consider SystemC itself as the high level system description language. However, at the application level one would like to have visual notations for interacting with the end users to capture requirements. It is also important to be able to use standard *models of computation* (MOCs) at the initial design stages. Further, one may not wish to concretely specify the communication mechanisms and instead leave it to be defined by the underlying operational semantics of the MOCs being deployed. Finally, design reuse with the help of modifications to an existing component as well as formal verification are easier to carry out at a higher level of abstraction than what is offered by SystemC. Hence we propose a top layer of system descriptions using UML notations.

Given these two choices, our goal is to build a flexible and automated translation mechanism using which one can transform UML based system descriptions to SystemC code. A crucial step here is to develop a *coherent subset* of UML notations. This is so because UML offers a bewildering variety of diagrammatic notations and it is up to the user to decide the combined roles of these various diagrams. We select here the so called executable subset of UML, namely class diagrams and state machine diagrams. The other diagram types may well be useful for capturing user requirements and for documenting important features of the design, but they are unlikely to contribute to code generation and hardware synthesis. One important exception is sequence diagrams. As we discuss later, they do have an important role to play in system level designs but we have not yet incorporated them in our framework.

The linkage between the UML layer and SystemC layer we have been constructing [211, 143] serves a dual purpose. On the one hand, we use it for transforming applications described at the UML layer to SystemC code for initial simulation. On the other hand, our translation mechanism also enables us to *pull up* the platform description mechanisms to the UML layer. In this latter usage, we could consider both the executable platform description and the application models to be available at the UML layer where one can hope to do formal verification. Further one can also begin to tackle a more abstract version of the problem of mapping an application to a platform. Using our translator, a designer can then translate these two descriptions down to the SystemC level for more detailed simulation and move towards a detailed implementation. With this as motivation, a substantial part of our work at the UML level consists of incorporating SystemC compatible (inspired) entities.

In our work , we have been mainly concerned with the *transaction level modeling* (TLM) layer of SystemC. At this level, the basic communication unit consists of a method call and hence the performance numbers reported will generally not be cycle accurate. This is acceptable if the goal is to rapidly obtain a design document at the SystemC level that describes both the application and the platform. Naturally, many other design steps will have to be realized to support a viable design flow, the key one among them being an efficient hardware synthesis tool. We feel that our modeling framework and the translator can easily be linked to tools that will provide these missing steps.

We have prototyped our UML based modeling environment using the Rhapsody 4.2 tool [172]. It supports state machine diagrams with concurrency and hierarchy (in other words, statecharts [85]). It also provides access to the XMI [155] representation of the design which is faithful and facilitates the translation process.

In order to support UML based platform descriptions, we have incorporated stereotypes in the Rhapsody environment to capture the communication primitives of SystemC such as interfaces and channels. On the hand, ports are declared as the internal attributes of modules and (hierarchical) channels. It is worth noting that the new version of Rhapsody [50] directly offers communication primitives with a similar flavor.

We have also incorporated the clock sensitivity features and other timing aspects of SystemC at the Rhapsody level. Consequently we can describe real time applications faithfully while being able to instrument performance constraints in the UML based platform descriptions.

As mentioned earlier, we use just class diagrams and state machine diagrams at present. More crucially, we only use the "unstructured" class diagrams and the restricted state diagrams of UML 1.5 while ignoring the simple sequence diagrams because they are not very useful for developing test benches. However, in UML 2.0, all three of these diagram types have been extended in powerful ways. A major challenge is to exploit these extensions while preserving the capabilities of our current framework. We shall return to this issue in Section 8.6.

8.1.1 Related Work

The need for system level design methods has been discussed more eloquently and in larger contexts in [124, 126, 80]. The role of SystemC in this context has also been explored in detail in [124, 80]. What UML may have to offer towards system level design methods for real time embedded systems has been studied from a number of perspectives as reported in [114]. For basic material on SystemC, the UML 2.0 standard and the Rhapsody tool, we refer the reader to [210, 80, 217] and [172, 50] respectively. Our programme, initiated here, could also have been based on system description languages such as SpecC, Rosetta or SystemVerilog [67, 3, 71]. Our preference for SystemC over these related languages has been mainly influenced by accessibility and familiarity. A similar remark applies to our choice of the Rhapsody tool.

UML and SystemC have also been proposed to be used elsewhere [221, 141, 26]. Our use of stereotypes for SystemC components is similar to those proposed in [221, 173, 26]. However, in these efforts, automatic generation of SystemC code from the models is not supported. An earlier effort that translates UML to SystemC is YAML [199]. However, YAML uses UML merely to capture the *structural* aspects of the system under design. In contrast, our approach provides for the full fledged use of state machine diagrams — including C++ code associated with the actions — and hence can capture *system behaviors* exhibiting concurrency at the UML level.

8.1.2 Organization of the Chapter

In the next section, we recall the main features of SystemC. In Section 8.3 we explain our scheme for using the Rhapsody tool to develop designs. In the subsequent sections we first discuss the major details of our implementation. We then present some examples and results to illustrate the main aspects of our translator. The final section concludes with a brief discussion on what needs to be achieved in order for UML 2.0 and SystemC to realize their potential in the domain of Systems-on-a-Chip design flows.

8.2 SystemC Preliminaries

Here we briefly describe the basic features of SystemC. For more information, please see [80] and [210]. SystemC is a library built entirely on top of C++. It separates computation and communication by having modules and processes for computation; ports, interfaces and channels for communication. Modules are the basic building blocks for partitioning a design. A module hides its data and algorithms from other modules. It may have one or more processes which can run concurrently. Modules as well as processes within a module communicate through channels. There are two types of channels: primitive channels and hierarchical channels. Primitive channels are in some sense, stateless while hierarchical channels can have internal states and control flow associated with them. As the name suggests, hierarchical channels can contain other channels, modules or processes. Interfaces specify the signature of the operations provided by channels. A module accesses a channel through a port whose type is one of the interfaces implemented by the channel.

A key feature of SystemC is that communication can be modeled at a high level of abstraction often referred to as *transaction level modeling* (TLM). It is hard to pin down this notion precisely. Intuitively, communication between components is described through method calls, without any synchronization. Here, 'transaction' stands for the exchange of data between two components of a system. This level emphasizes what data are transferred and from which locations but not the details of the specific protocol used by the communication. Thus, intercomponent interactions are abstracted from the details of the implementation of the communication architecture and this facilitates component reuse. In addition, simulation at this level can be usually carried out at a much high speed. For a more detailed description of TLM, see [30].

Behavioral synthesis of SystemC descriptions is still an unfinished story. On this front, one tool we were able to access is the CoCentric SystemC Compiler tool. It synthesizes a SystemC behavioral hardware module into an RTL description or a gate level netlist. Unfortunately, severe restrictions are placed on the SystemC code that can be synthesized. Further, this tool is no longer supported by its vendor. More recently, two new synthesis tools [32, 63] have come to the market but it is too early to assess their strengths and weaknesses.

8.3 UML Modeling

We use two types of UML diagrams for modeling, namely, class and state machine diagrams. Class diagrams are mainly used to describe the component structure of a system while state machine diagrams describe the behavior of the components. Besides the standard UML notations, we have also lifted some SystemC features up to the UML level using the stereotype extension mechanism.

8.3.1 Class Diagrams

We use the class hierarchy in the usual way to describe the computational entities via their methods and data types. More crucially, class diagrams are also used to specify the overall structure of a system in terms of its components and how the components are connected to each other. We wish to emphasize that we do not handle at present the structured classes notation offered in UML 2.0. This is a rich extension particularly relevant for describing complex hierarchical multicomponent systems. Augmenting the current version of our translator to incorporate structured classes will considerably expand its applicability. We shall return to this issue in the Section 8.6.

To bring out the main aspects of our modeling method, we will use the simple bus model available in the SystemC package [210] as a running example. In this system, there are three masters, namely a blocking, a nonblocking and a direct master. In addition, there is a bus and two memory slaves. A master initiates transactions on the bus to access a memory. Figure 8.1 shows a fragment of the class diagram of this example.

Class notations are also used to define and distinguish between the various features of SystemC lifted up to the UML level using the stereotype mechanism. This is an extension mechanism of UML that allows one to define virtual sub-

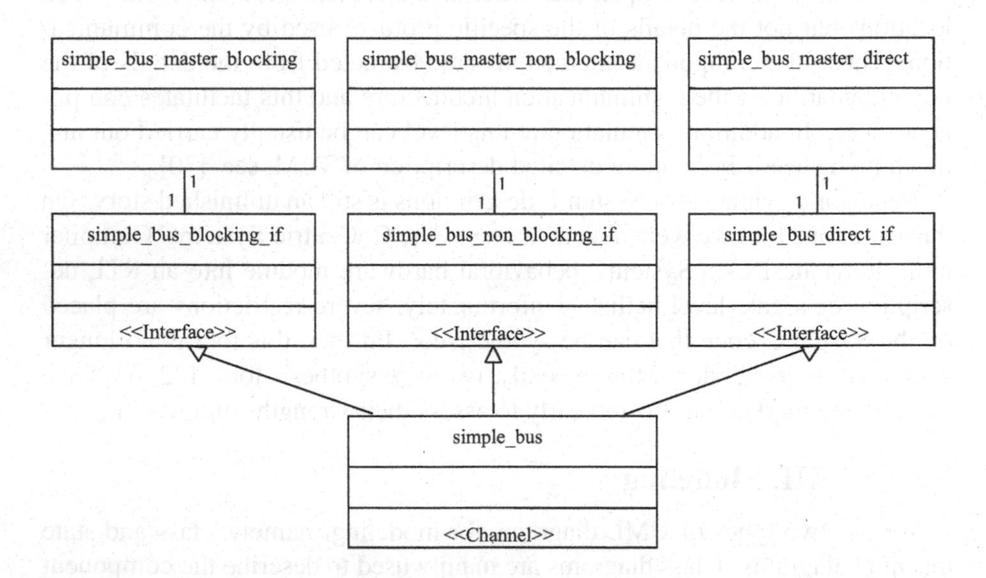

Figure 8.1. A Class Diagram

classes of UML meta classes with new meta attributes and additional semantics. Using this, users can define a class as a module, an interface, a primitive channel or a hierarchical channel. In addition, we support declaration of ports for modules to access channels. A port can be declared as an attribute of a module.

Classes can be related by the following UML compatible relations:

- Generalization (or inheritance): when a channel implements an interface, it inherits that interface. Moreover, an interface, channel and module can inherit another interface, channel and module respectively.
- Aggregation/composition: modules and channels may be hierarchical.
- Association: classes that exchange messages with each other are associated to one another. We model messages by UML events with or without arguments. Furthermore, a module may have an association relationship with an interface when it accesses a channel through this interface.

In the case of the simple bus example, the masters access the bus through three different interfaces. The bus is a hierarchical channel which implements the three interfaces.

8.3.2 State Machine Diagrams

State machine diagrams describe the behavior of classes. A state can be a simple state or a composite state. A composite state may consist of concurrent substates; in this case it is called an orthogonal state. A composite state which consists of sequential substates is called a simple composite state. Being in an orthogonal state means being in all of its substates. Being in a simple composite state means being in exactly one of its substates.

Modeling concurrency is an important part of a system specification and this is achieved with the help of orthogonal states. Figure 8.2 shows a state machine diagram of a master which is a combination of the three masters described above. This is a derived version of the simple bus model in the SystemC package; we have combined the three masters into one master state machine diagram. The orthogonal state `Master` has three substates, each of which is a simple composite state which in turn has a set of simple leaf states (that have no internal structure). Each state is associated with a set of actions on entry and actions on exit. These will be executed when the object enters and leaves that state respectively. A transition connects a source state and a target state. The label of each transition includes a trigger event, a guard and a sequence of actions. Events may be parameterized. A guard is an expression which returns a Boolean value. When an object is in a state and an event of an outgoing transition of that state occurs, the corresponding guard is evaluated. If the guard is true, the transition is taken, the actions are performed and the object

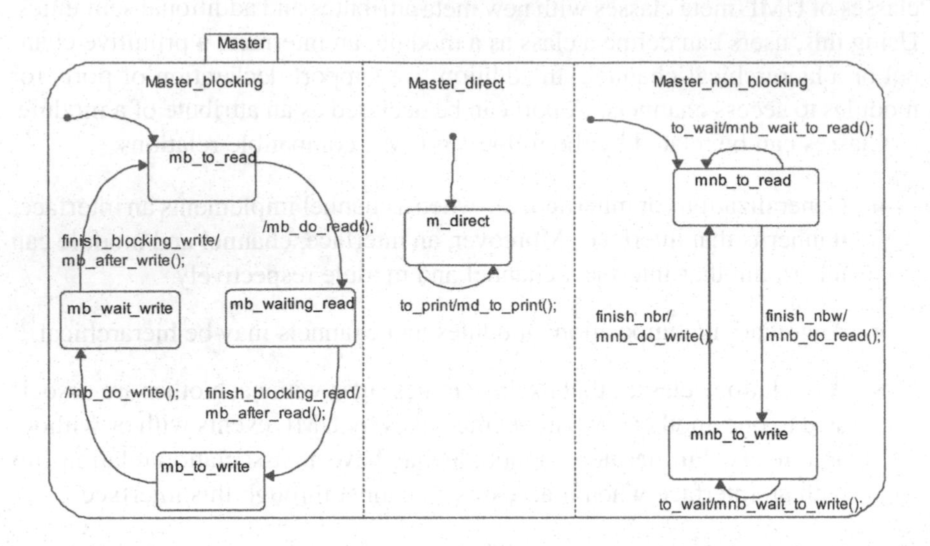

Figure 8.2. An Orthogonal State

moves into another state. Otherwise, the object stays in that state and the event is simply discarded.

The actions associated with a transition can be C++ statements or a function call whose body (in the form C++ code) is to be provided by the user. The action could also correspond to sending an event to another state machine diagram (describing the behavior of a different class). In addition, the action could be calling an interface method through a port. Moreover, in the actions, we support specification of clock sensitivity or delays in terms of clock cycles or time units through C++ macros. This gives the designers an option to have timed models. Furthermore, this allows users to provide annotations of timing information for performance estimation and architectural exploration. For TLM level implementations, we do not restrict the C++ code associated with the actions in anyway. There will however be severe restrictions when the target is behavioral level SystemC code. We will return to this point in the next section.

Figure 8.2 shows an orthogonal state named `Master` consisting of three states: `Master_direct`, `Master_blocking` and `Master_non_blocking`. We describe the behavior associated with the `Master_blocking` state. First, it goes from the initial state to `mb_to_read` state. Since there is no trigger event and guard for the transition, the function `mb_do_read` is called and the state `mb_waiting_read` is entered. In state `mb_waiting_read`, when the event `finish_blocking_read` arrives, `mb_after_read` is performed and state `mb_to_write` is entered. Other states and transitions can be interpreted similarly.

In our framework, classes or objects can communicate through events that carry arguments. On the other hand, classes corresponding to modules or hierarchical channels may also communicate by means of interface methods.

We currently require users to declare a top level class to instantiate objects. Object diagrams could have been used instead to do this. We also allow for only one level of nesting in our state machine diagrams with orthogonal states at the top layer and simple composite states at the second layer. However it will not be difficult to extend our state machine diagrams to allow more than two levels of hierarchy.

8.4 Implementation

As mentioned earlier, the Rhapsody tool supports the main features of UML that we need. Moreover, Rhapsody has a toolkit which can generate XMI [155] as an intermediate representation. This representation contains all the information about the model that we need for code generation.

The XMI toolkit is used to generate XMI document from the graphical models. We then use our XMI parser to gather information from the XMI document to build an abstract tree as an input to a template engine called Velocity [224] that generates SystemC code from predefined templates. With the help of this engine, we are able to decouple the parsing of XMI document from the code generation step so that changes in the XMI parser do not affect the code generation process. Further, in the later part of the work flow, we only need to work with the templates to generate code without having to deal with the verbose code of the parser. Consequently, by merely modifying the templates for one level of abstraction, we can have the templates for another level of abstraction and generate code, without touching the XMI parser. Figure 8.3 shows the work flow of our translator.

We support the initialization of multiple instances of a type (module, primitive channel or hierarchical channel). However, they cannot be created dynamically since SystemC does not support dynamic instantiation; the structure of a system is determined at elaboration time.

The skeleton of the SystemC modules, interfaces, channels and their relationship are generated from class diagrams in a straightforward fashion. The action code for modules and channels are generated from their state machine diagrams.

From the model that includes the state machine diagram in Figure 8.4, an XMI document is generated from the Rhapsody XMI tool kit. Here is a fragment of the rather verbose XMI code that describes the transition from state `s1` to state `s2`. It is not meant to be readable. We are including it merely for the sake of completeness.

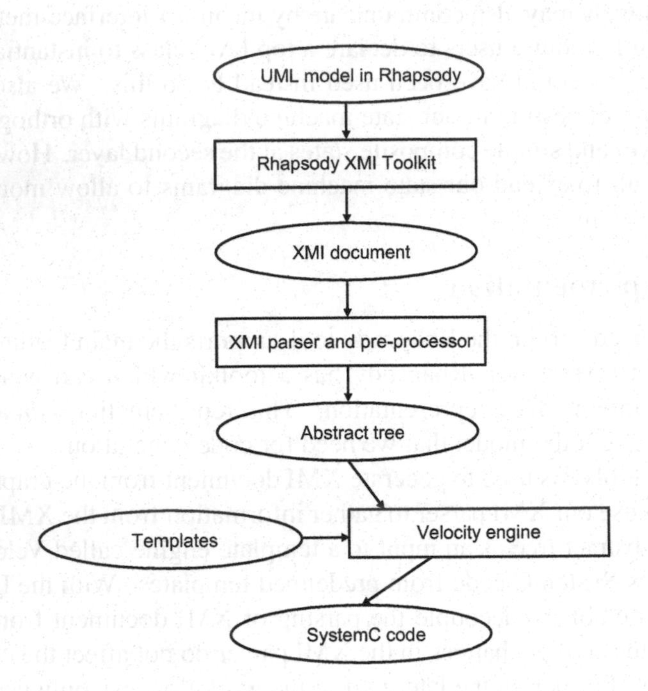

Figure 8.3. Our Implementation Workflow

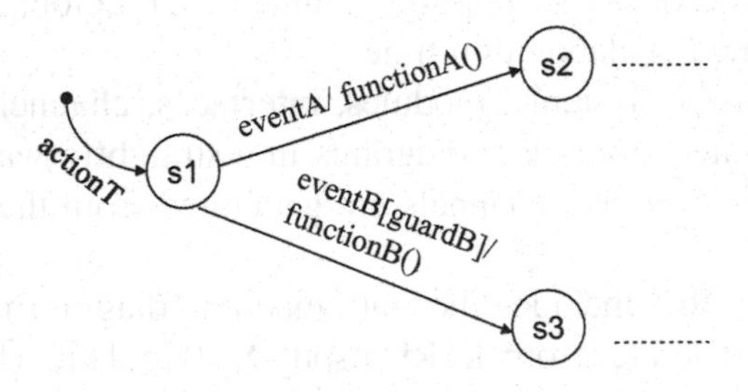

Figure 8.4. A Simple State Machine Diagram

```
<Behavioral_Elements.State_Machines.Transition xmi.id="_52"
xmi.uuid="GUID 4b07ad6f-53cf-4f08-b181-3f1ad9b809a2">
    <Foundation.Core.ModelElement.name>2</Foundation.Core.
                                            ModelElement.name>
    <Foundation.Core.ModelElement.visibility xmi.value="public"/>
    <Behavioral_Elements.State_Machines.Transition.trigger>
        <Behavioral_Elements.State_Machines.SignalEvent xmi.idref="_55"/>
        <!-- call eventA -->
    </Behavioral_Elements.State_Machines.Transition.trigger>
    <Behavioral_Elements.State_Machines.Transition.source>
        <Behavioral_Elements.State_Machines.State xmi.idref="_53"/>
        <!-- s1 -->
    </Behavioral_Elements.State_Machines.Transition.source>
    <Behavioral_Elements.State_Machines.Transition.target>
        <Behavioral_Elements.State_Machines.State xmi.idref="_49"/>
        <!-- s2 -->
    </Behavioral_Elements.State_Machines.Transition.target>
    <Foundation.Core.ModelElement.taggedValue>
        <Foundation.Extension_Mechanisms.TaggedValue>
            <Foundation.Extension_Mechanisms.TaggedValue.tag>
                                                    displayName
            </Foundation.Extension_Mechanisms.TaggedValue.tag>
            <Foundation.Extension_Mechanisms.TaggedValue.value>
                eventA/functionA();
            </Foundation.Extension_Mechanisms.TaggedValue.value>
        </Foundation.Extension_Mechanisms.TaggedValue>
    </Foundation.Core.ModelElement.taggedValue>
    <Behavioral_Elements.State_Machines.Transition.effect>
        <Behavioral_Elements.Common_Behavior.UninterpretedAction
        xmi.id="_56" xmi.uuid="GUID 8ac5ea7f-6e32-4151-8530-16712de444a7">
            <Foundation.Core.ModelElement.name>
                functionA();
            </Foundation.Core.ModelElement.name>
            <Foundation.Core.ModelElement.visibility xmi.value="public"/>
            <Behavioral_Elements.Common_Behavior.Action.script>
                <Foundation.Data_Types.ActionExpression xmi.id="_57">
                    <Foundation.Data_Types.Expression.body>
                        functionA();
                    </Foundation.Data_Types.Expression.body>
                </Foundation.Data_Types.ActionExpression>
            </Behavioral_Elements.Common_Behavior.Action.script>
        </Behavioral_Elements.Common_Behavior.UninterpretedAction>
    </Behavioral_Elements.State_Machines.Transition.effect>
</Behavioral_Elements.State_Machines.Transition>
```

After the XMI parser has read the XMI documents that contains code such as the one shown above, it builds an abstract tree. The Velocity engine will then take this tree and together with the necessary templates as its input and generates SystemC code. It is the templates that determine how the generated SystemC code should look like. We show here a simplified fragment of a template that is used for the state machine diagrams.

```
   switch (state)
   {
#foreach ($state in $class.getRootState().getInterStates())
   case $state.getID():
      $state.getActionOnEntry()
      this_uport->wait#foreach($transition in
      $state.getOutgoingTransitions())_$transition.getTrigger()#end();
#foreach ($transition in $state.getOutgoingTransitions())
      if (this_uport->get_$transition.getTrigger()_flag() == true)
      {
         if ($transition.getGuards())
         {
            $state.getActionOnExit()
            $transition.getAction()
            state = $transition.getTarget();
            break;
         }
      }
   }
#end
#end
```

The following is the SystemC code which will be generated for the simple state machine diagram in Figure 8.4. For the sake of readability we present this program in pseudocode form.

```
current_state = initial_state;
while (current_state != final_state) {
   switch(state) {
   case initial_state:
      wait for trigger from the top module to start behavior
      actionT;
      current_state = s1;
      break;
   case s1:
      actions on entry of s1
      wait for eventA or eventB to come
      if eventA comes {
         if (true) { // the guard of this transition is true
            actions on exit of s1
            functionA();
            current_state = s2;
            break;
         }
      }
      if eventB comes {
         if guardB is true {
            actions on exit of s1
            functionB();
            current_state = s3;
            break;
         }
      }
      break;
   case s2: . . . break; case s3: . . . break;
   }
}
```

The code of `actionT`, `functionA` and `functionB` are to be provided by the users. The SystemC implementation of actions for events depends on the levels of abstraction which will be discussed in the following subsections.

8.4.1 Translation to TLM Level

SystemC code at the TLM level is ideal for simulation as details of the low level communication infrastructure are not present. In our design flow, users do not have to specify any SystemC components at UML level. They can simply work with classes or objects, state machine diagrams and model communication between objects by events with or without arguments. Such events will be implemented through SystemC primitive channels. Each module has a primitive channel to receive events sent by other modules. This primitive channel essentially acts as a buffer for incoming events to that module. When a module has an association relationship with another module, it can send messages to that module. A port is declared in the sender module to access the other module's primitive channel. Thus, the primitive channels implement two interfaces: the interface for sending events and the interface for receiving events. The SystemC code generated by our translator will be at the TLM level since the senders and receivers just call functions of the primitive channels, regardless of whether or not the events have arguments associated with them. In the pseudocode above, waiting for an event to arrive or checking if some event has arrived constitutes function calls to the corresponding primitive channels. Users can also specify SystemC components, such as interfaces and channels through stereotypes, ports, time delay, and clock sensitivity through C++ macros. These are translated to SystemC accordingly.

There are three types of SystemC processes: `sc_thread`, `sc_cthread` and `sc_method` [80]. Each simple composite state in a state machine diagram is translated into a `sc_thread` of the corresponding module. Thus, a state machine diagram such as the one in Figure 8.2 will be translated into three `sc_threads` of the same module. The reason we chose `sc_thread` over `sc_method` and `sc_cthread` is that an `sc_thread` can be suspended during execution to wait for events and is not necessarily sensitive to every clock edge.

8.4.2 Translation to Behavioral Level

We have also experimented with the generation of behavioral level SystemC code using CoCentric tool of Synopsys. This tool requires us to place rather severe restrictions at the Rhapsody level on the C++ code fragments supplied by the user. Further, one has to declare a class called `Top` to initialize all the instances since the method `new()` used to create instances is not synthesizable. One may however initialize multiple instances of the same class. The translator will create the corresponding modules and connect them according to their specified relationships.

The code synthesized at this level has to comply with the coding convention of the Synopsys tool. Restrictions are placed on the data types, constructs, instructions and SystemC classes [208]. Due to these restrictions, a simple composite state is translated to an `sc_cthread` which is only sensitive to an active clock edge. Further, communication is achieved only through signals. UML events are implemented as `sc_signals` which can have boolean values representing the existence of the events. We will see an example of the behavioral level code in the next section.

8.5 Examples and Results

8.5.1 A Simple Bus

This is a benchmark example of SystemC at the TLM level which has been described partially in the previous section. Here we use it to demonstrate how one might model a fragment of a platform at UML level and translate it into SystemC. This model uses all the four stereotypes mentioned above, namely modules, interfaces, primitive channels and hierarchical channels. In particular, the three masters are modules that access the bus through three ports using three different interfaces. The bus is a hierarchical channel which implements the methods of the three interfaces.

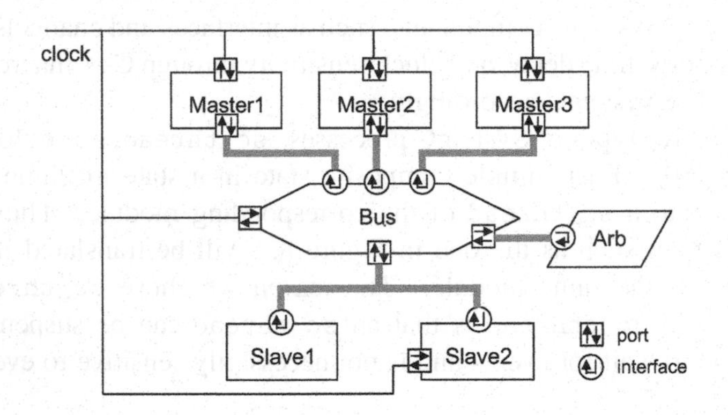

Figure 8.5. Block Diagram for the Simple Bus Example [80]

For faster simulation speed, the arbiter and the fast memory are modelled as primitive channels to decrease the number of threads and thus, decrease the context switching time. The bus accesses these primitive channels through the arbiter interface and the slave interface, respectively. The `Top` class initializes all the objects of the system, in this case one instance for each module and channel. Figure 8.6 shows the class diagram of this example.

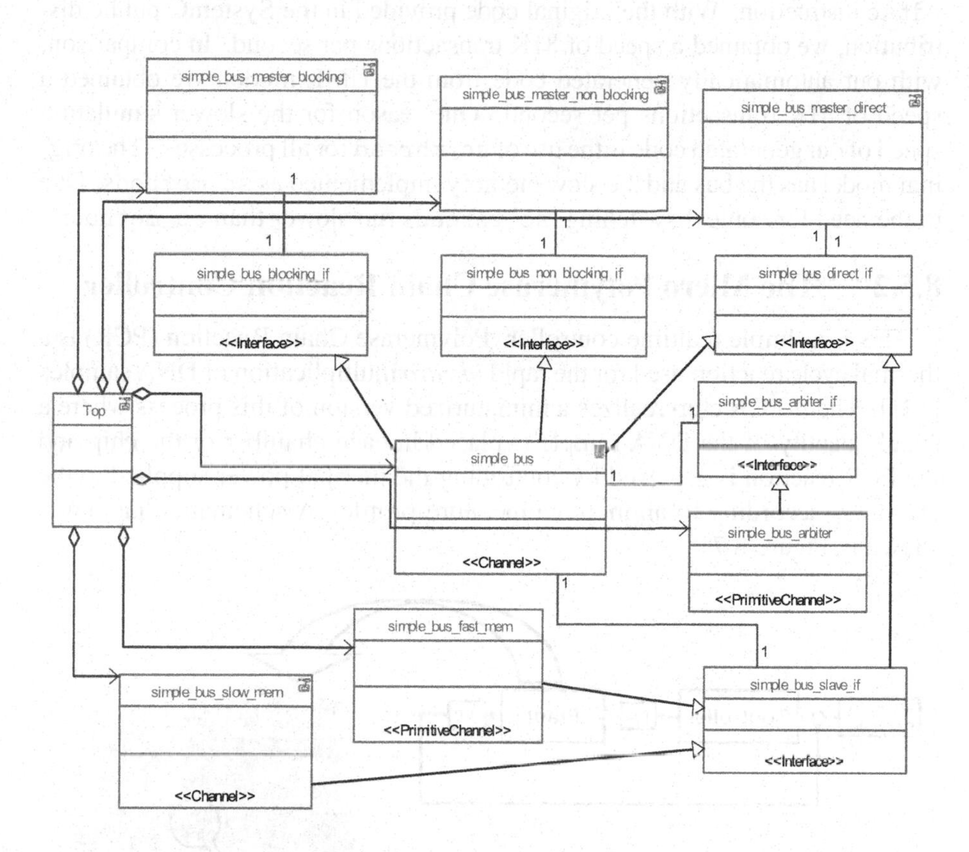

Figure 8.6. Class Diagram of the Simple Bus Example

We first captured this UML level model using the Rhapsody tool and then translated it into TLM SystemC code automatically using our translator. We then simulated the resulting SystemC code using the standard SystemC simulation kernel. When measuring performance, we did not initialize the direct master, because it is only used for debugging. The experiments were performed on Linux Red Hat 9.0 running on CPU Intel Xeon 2.8GHz. We measured the number of CPU clock cycles for 1,000 bus transactions using the Pentium's `rdtsc` instruction. With the original code provided in the SystemC public distribution, we obtained a speed of 81K transactions per second. In comparison, with our automatically generated code from the UML model, we obtained a speed of 41K transactions per second. One reason for the slower simulation speed of our generated code is the use of `sc_thread` for all processes. The original model has the bus and the slow memory implemented as `sc_methods`. Due to the need for context switching, `sc_threads` run slower than `sc_methods`.

8.5.2 The Micro Polymerase Chain Reaction Controller

This is a simple realtime controller. Polymerase Chain Reaction (PCR) is a thermal cycle reaction used for the rapid *in vitro* multiplication of DNA samples [121]. The μ-PCR chip realizes a miniaturized version of this process where a small quantity of the DNA sample is placed in each chamber of the chip and the PCR reaction is achieved by controlling the thermal power supplied to the chambers according to an input temperature profile. A schematic diagram is shown in Figure 8.7.

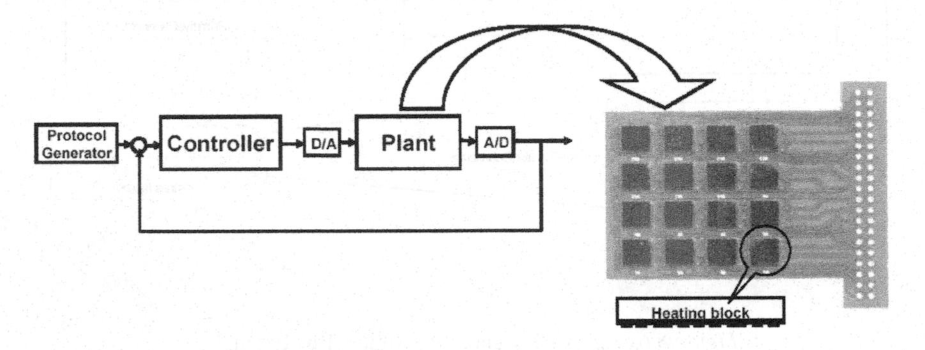

Figure 8.7. μ-PCR Block Diagram

We will not describe here the PCR biochemical process in detail but instead focus on the functional model of the controller. This unit is driven by the temperature profile, which specifies the control objective, and feedback received from the chip regarding the current temperatures of the chambers. In the present version of the plant model, the effects of interchamber influences are ignored as a simplification. Hence there is one independent controller for each chamber.

This controller periodically reads the temperature, converted into a voltage value via an analog-to-digital converter, of the chamber. With the help of the estimator — the control law — it then computes the output voltage required for that chamber to maintain the temperature according to the temperature profile of the current PCR thermal cycle. This voltage is then converted back into an analog value via a digital-to-analog converter, which is then used to control the heating element of that chamber.

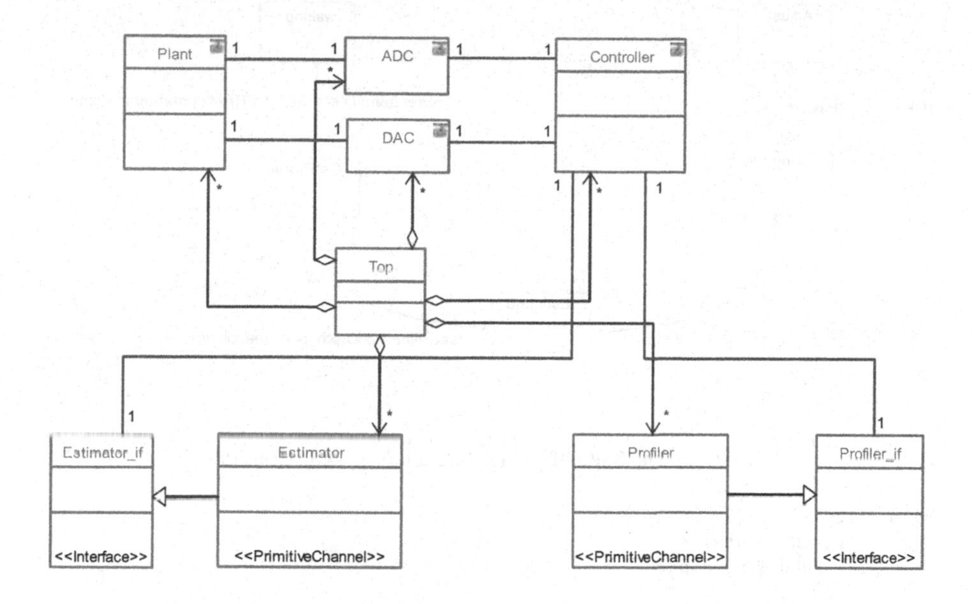

Figure 8.8. μ-PCR Class Diagram

Figure 8.8 shows the class diagram of this example. The profiler that keeps the temperature profile and the estimator that keeps the control laws were modelled separately from the controller so that we can reconfigure the temperature profile and control law easily. They were modelled as primitive channels in order to get better simulation speed at the TLM level. The communication among the modules are cycle accurate, in the sense the status of a module's input and output are specified at each clock cycle. Yet another real time aspect is the timing diagram associated with the A/D converter. The state machine diagram of the controller is shown in Figure 8.9.

For this example, we have synthesized, using the CoCentric compiler tool of Synopsys, the behavioral level SystemC code generated via our translator. This application has been simulated at both TLM and behavior levels.

Following is a fragment of the code for the `sc_cthread` of the controller at behavior level. For brevity we have eliminated some `wait()` statements.

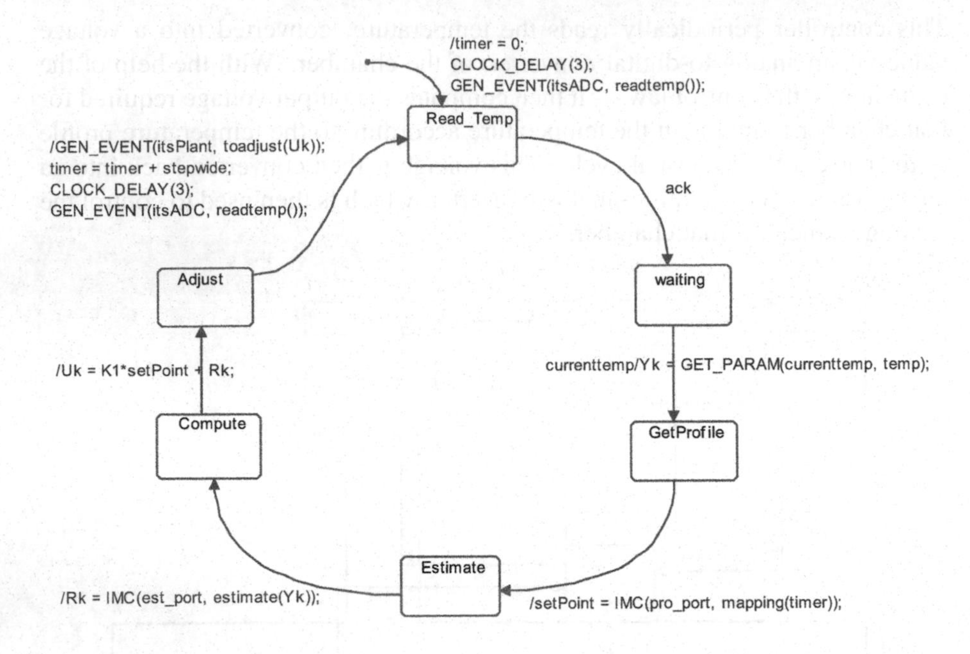

Figure 8.9. The State Machine Diagram of μ-PCR Controller

```
while (true) {
        switch (state) {
        // Read_Temp state
        case 106:
                wait_ack(); //wait for ack signal from ADC
                if ( read_ack() == true ) {
                   //the guard of this transition is true
                   if (true) {
                      state = 118;
                      write_ack(false);
                      break;
                   }
                   else
                      write_ack(false);
                }
                break;
        //GetProfile state
        case 110:
                // no event trigger for this transition
                if (true) {
                   //IMC is a macro for an interface method call
                   //to the port pro_port to the profiler
                   //the method mapping has an argument timer
                   //the returned value of this method
                   //is assigned to variable setPoint
                   setPoint = IMC(pro_port, mapping(timer));
                   state = 115;
                }
```

```
            break;
     //Adjust state
     case 113:
            if (true){
               GEN_EVENT(itsPlant, toadjust(Uk));
               timer = timer + stepwide;
               //this macro is called to delay for 3 clock cycles
               CLOCK_DELAY(3);
               //send event readtemp to the ADC
               GEN_EVENT(itsADC, readtemp());
               state = 106;
            }
            break;
     //Estimate state
     case 115:
            if (true) {
               Rk = IMC(est_port, estimate(Yk));
               state = 117;
            }
            break;
     //Compute state
     case 117:
            if (true){
               Uk = K1*setPoint + Rk;
               state = 113;
            }
            break;
     //Waiting state
     case 118:
            wait_currenttemp();
            if (read_currenttemp() == true){
               if (true){
                  //GET_PARAM is the macro to get
                  //the value of parameter temp
                  //associated with currenttemp
                  //event
                  Yk = GET_PARAM(currenttemp, temp);
                  state = 110;
                  write_currenttemp(false);
                  break;
               }
               else
                  write_currenttemp(false);
            }
            break;
     // the initial state
     case 119:
            //wait for signal to start the execution
            //of this process
            wait_initController();
            timer = 0;
            GEN_EVENT(itsADC, readtemp());
            CLOCK_DELAY(3);
            state = 106;
            break;

     }
}
```

The implementation of the functions like `mapping()` and `estimate()` is provided by users; they can only use the synthesizable subset of SystemC defined by Synopsys. Note that although SystemC interfaces and channels are not synthesizable by CoCentric Synopsys, we can still generate behavioral level SystemC code from the models that have interfaces and channels like the one in this example. In this case, the interfaces in the UML models are not generated, the model's channels are declared as `sc_modules`; each has a thread that receives triggers for method calls from other modules, locally calls the methods and returns values by sending signals to the caller modules.

Table 8.1 shows the simulation speed—in terms of transactions per second—of the μ-PCR example on the same platform as the one used in the previous example. By a transaction we mean the period of time in which the controller senses the current voltage, computes and outputs to the plant.

Our simulation results show, as expected, that simulation speed at the TLM level is higher than that at the behavior level. The experiments also give evidence that the code we generate scales fairly well in terms of performance.

Table 8.1. Simulation Speed of the μ-PCR Example

Chamber arrangement	TLM sim. (trans./sec)	Behavioral sim. (trans./sec)
2×2	12,125	5,766
4×4	2,714	1,543
5×5	1,676	785
8×8	555	148

8.5.3 Digital Down Converter

For our third example, we implemented a *digital down converter* (DDC) for the *global system for mobile communications* (GSM) - a wireless communication protocol . Digital radio receivers often have fast analog to digital converters delivering vast amounts of data. However, in many cases, the signal of interest represents a small proportion of that bandwidth. A DDC is a filter that extracts the signal of interest from the incoming data stream. Our implementation closely follows the MATLAB example in Xilinx's system generator (see Figure 8.10).

The desired channel is translated to baseband using the digital mixer comprised of multipliers and a direct digital synthesizer (DDS). The sample rate of the signal is then adjusted by a multistage, multirate filter consisting of a cascade integrator-comb (CIC) filter and two polyphase finite impulse response (FIR) filters with a decimation factor of 2. The functions performed in the system are complex multiplication and multirate filtering. The overall down sampling rate of the converter is 192:1.

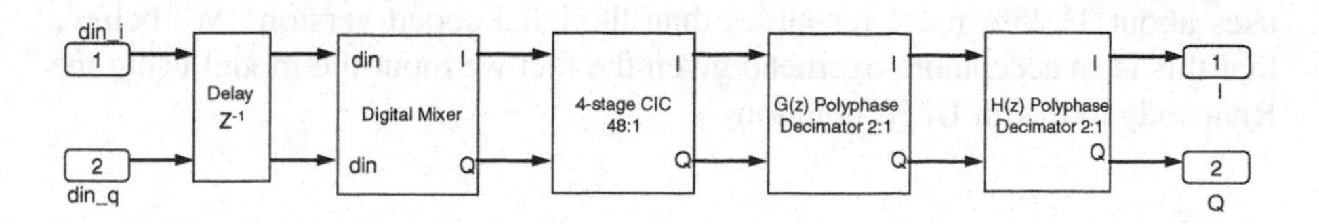

Figure 8.10. Digital Down Converter for GSM - Block Diagram

Each of the components is mapped to a module, and data is sent through the chain by events (see Figure 8.11). The model has been translated into both TLM and behavioral levels. We could not find the source code for a similar DDC in UML or SystemC for comparison. Hence we could only compare the FIR module of our design with an FIR example provided by Synopsys. The only modification we did to the Synopsys code was to ensure that the coefficients and the bit widths of the ports are the same as those of our FIR model. The codes were compiled into gate level netlist using Synopsys `tc6a_cbacore` library, which targets cell based array architectures [209]. The same timing constraints were used on the synthesis runs of both. Table 8.2 shows the comparisons of the final synthesized hardware. From the result we can see that our generated code

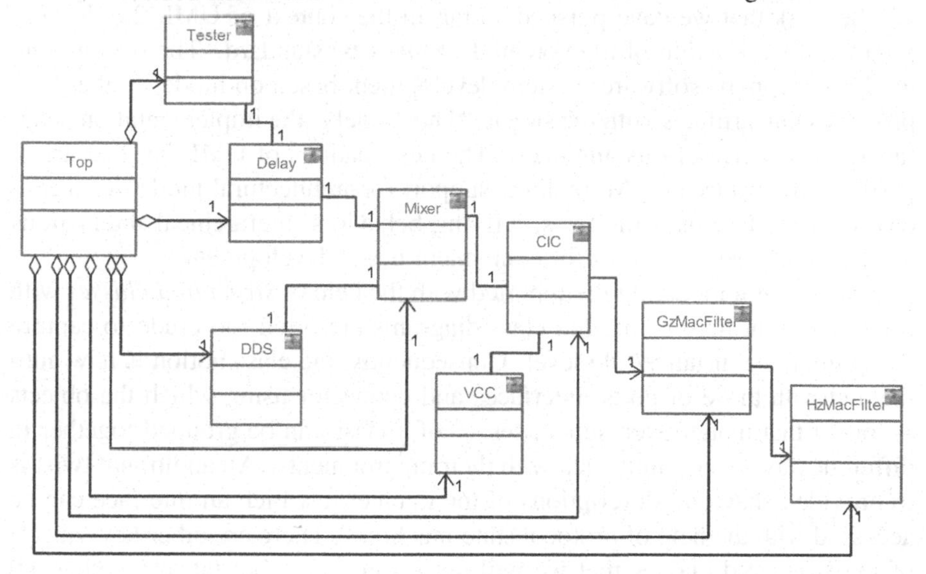

Figure 8.11. Digital Down Converter for GSM - Class Diagram

uses about 33.25% more resources than the hand coded version. We believe that this is an acceptable overhead given the fact we input the model using the Rhapsody tool with UML notations.

Table 8.2. Area Statistics for FIR Component Implemented on Cell Based Array Architecture

	FIR from Synopsys(S)	FIR from DDC(D)	Ratio((D-S)/S)
Number of ports	260	261	0.39%
Number of nets	18393	27942	51.92%
Number of cells	18010	27547	55.15%
Number of references	93	99	6.45%
Combinational area	30181.2	50583.7	67.60%
Noncombinational area	34560.0	36844.2	6.61%
Net interconnect area	244806.2	325033.1	32.77%
Total cell area	64741.1	87430.3	35.05%
Total area	309547.6	412461.1	33.25%

8.6 Conclusion

The work that we have pursued so far, in the context of UML 2.0, has exploited only a fraction of the potential of this new standard. The overarching goal is to support (software) system development based on models rather than programming artifacts with the slogan "The model is the implementation" capturing the essence of this approach. The new features of UML 2.0 that are of particular relevance are: More direct support for architectural modeling, a generous menu of formalisms for specifying behaviors, hierarchical interactions modeling and better support for component based development.

Architectural modeling is supported with the help of *structured classes* with the key insight here being that class diagrams are often too crude to capture structure at the instance/role level. Consequently, the class notion is now augmented with those of ports, interfaces and connector using which the objects — or for that matter even subclasses — of a class can be grouped together in different ways to communicate with their environments. An additional twist is to provide behavioral descriptions of the manner in which an interface can be accessed with the help of protocol state machines. There are other key aspects of the structured classes that we will not go into here but taken together, all these features make structured classes into a powerful mechanism for specifying complex, hierarchical, component based architectures.

State machine diagrams, the basic behavioral specification mechanism, have also been enhanced in UML 2.0 with the features of modularized submachines and specialization/redefinition, action blocks and state lists.

Yet another key feature of UML 2.0, from the standpoint of system level design are the sequence diagrams which can now be structurally decomposed. In addition they can be composed with help of operations such as alternatives, iteration, break (to exit from a loop), negative (forbidden scenarios) and conditions. In addition, timing diagrams are also a part of the interactions classifier. As a result, the designer can now develop powerful test benches along with the system specification. As a result, both the specification and its test benches can be compiled into SystemC code for simulation and verification.

In summary, structured classes, state machine diagrams, protocol state machines, sequence diagrams and timing diagrams of UML 2.0 together constitute a powerful conceptual and notational base for developing system level designs. The key to realizing this potential is to *automatically* generate *executable* code from specifications developed using these diagram types so that one can carry out simulation, performance estimation and verification. We feel that the UML-SystemC bridge that we are advocating here can, with concerted effort, help achieve this purpose.

We have presented here the backbone of a framework in which designs can be specified using UML notations. SystemC code implementing these designs can then be automatically generated. We showed some realistic examples illustrating the use of object oriented structuring, real time constraints and transaction level modeling. We see this framework as a sound launching pad for realizing the considerable potential that UML 2.0 has to serve as the basis for model driven system design methods.

Chapter 9

A Comparison between UML and Function Blocks for Heterogeneous SoC Design and ASIP Generation

Lisane Brisolara,[1] Leandro B. Becker,[2] Luigi Carro,[3] Flavio Wagner,[1]
Carlos E. Pereira[3]

[1]*Computer Science Institute*
Federal University of Rio Grande do Sul (UFRGS)
Porto Alegre, Brazil

[2]*Automation and Control Systems Department*
Federal University of Santa Catarina (UFSC)
Florianópolis, Brazil

[3]*Electrical Engineering Department*
Federal University of Rio Grande do Sul (UFRGS)
Porto Alegre, Brazil

Abstract This chapter presents the SEEP methodology for heterogeneous SOC design and ASIP generation starting from high level models. However, the main point is the comparison between the Functional Blocks and the Unified Modeling Language modeling approaches. Results obtained have led to the use of UML within the scope of SEEP methodology. Related issues are discussed in detail, together with a case study.

9.1 Introduction

The great majority of current electronic products contain an embedded computational system, for example mobile telephones, DVD players, microwaves ovens, etc.. Moreover, embedded controllers are also used in systems whose purposes are far from the electronic domain, such as medical devices, cars, and airplanes. Such embedded computational systems are often implemented as heterogeneous systems-on-a-chip (SoCs), which are usually composed of ded-

G. Martin and W. Müller (eds.), UML for SOC Design, 199–222.

icated hardware modules, programmable processors, memory, interface controllers, and software components.

Embedded systems design has become an important research area owing to the high complexity of new generations of systems. Increasing complexity derives from the amount of functionality that is associated with those systems. Efforts in all areas of system level design, such as specification, modeling, synthesis, simulation, verification, and estimations are required in order to cope with this increasing complexity.

At the same time the life cycle of embedded products becomes increasingly tighter. Cell phones, for example, represent one of many examples of this trend, since they are the basis for two major product line developments each year, compared to only one a few years ago [105]. In this scenario productivity and quality are simultaneously required in embedded systems design in order to deliver competitive products. Current research on embedded systems design emphasizes that the use of techniques starting from higher abstraction levels is crucial to the design success. Some authors such as Selic [191], Douglass [49], and Gomaa [72] argue that this approach is the only viable way of coping with the complexity which is found in the new generations of embedded systems.

Using this approach, models of embedded systems should evolve from high level views into actual implementations, ensuring a relatively smooth and potentially much more reliable process in comparison with traditional forms of engineering. Thereby the Unified Modeling Language (UML) has gained in popularity as a tool for specification and design of embedded systems and SoCs. In [114] one can find several efforts that describe the use of UML during the different phases of an embedded system design process. Such popularization comes from UML being by far the most used modeling notation for conventional computational systems.

Another issue is that owing to their various applications many embedded systems can be considered heterogeneous from the problem domain perspective. This applies to systems whose respective models require different models of computation (MoCs) [56], such as stream processing, control flow, and continuous time, in order to capture and express their behavior. For example, the specification of a mobile phone requires not only digital signal processing for the telecommunication domain, which is a discrete time MoC, but also sequential logic programs to describe several available applications (e.g., contacts and alarm clock). Such requirements contrast with the characteristics of UML, which was originally designed for the specification of event based systems.

Traditionally, the functional block (FB) modeling approach has been used by the signal processing, industrial automation, and control engineering communities for the development of embedded systems (see [103]). These models are widely accepted in industrial design, driven by an extensive set of design tools, such as, for instance, Matlab/Simulink from MathWorks. Features such as modularity, abstraction level, and re-usability contributed to the popularity of this modeling approach.

A relevant question is whether the use of UML presents concrete advantages for the design of embedded systems (further implemented as a SoC), when compared to a more traditional approach such as the FB approach. To answer this question a qualitative and quantitative comparison between both approaches is presented in this chapter by means of a case study. Results show that by using a specific toolset it is possible to generate a SoC from the UML model. Moreover, the current weaknesses of UML for SoC generation are highlighted and possible solutions are discussed.

This chapter introduces the Object Oriented Platform Based Design Process for Embedded Real Time Systems, or simply SEEP (from the Portuguese acronym), which proposes a methodology for SoC design and ASIP generation. This methodology offers a complete set of modeling, analysis, validation, and synthesis tools to support the development of optimized embedded real time systems comprising software and hardware components. It is based on the re-use of hardware and software components and on the configuration of architectural platforms implemented upon affordable FPGAs.

The remaining parts of the chapter are divided in the following way. Section 9.2 presents the SEEP methodology for SoCs design and ASIP generation. Section 9.3 presents the case study, the developed models, as well as the results from the comparison between UML and FB. Section 9.4 discusses weaknesses of UML that must be further tackled to allow the complete generation of SoCs. Section 9.5 discusses related work. Lastly, the main conclusions of this chapter are drawn in Section 9.6.

9.2 A Methodology for SoC Design and ASIP Generation

The SEEP methodology proposes a complete and integrated approach for SoC design and ASIP generation. This methodology is defined to guide the system integrator and the core provider towards the development of embedded applications within a reduced design time. Therefore the re-use concept is assumed, and each design step aims at facilitating the development of re-usable components and the rapid, but cost effective, design space exploration. The proposed methodology is basically divided in two phases, as follows: (i) architecture-independent design; and (ii) system generation.

An overview of the proposed design methodology flow is shown in Figure 9.1. The first phase generates as output the so called architecture-independent specification, which is further used as input for the second phase which generates the ASIP.

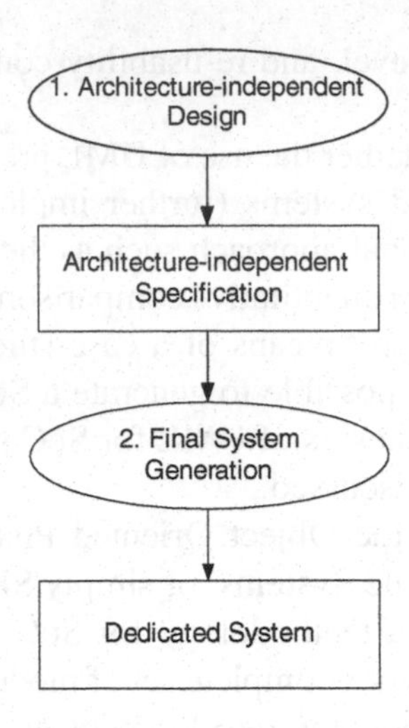

Figure 9.1. Overview of SEEP Methodology for SoC Design and ASIP Generation

9.2.1 The Architecture-Independent Design

This phase is responsible for generating the architecture-independent specification, which can be understood as the source code of the application that contains the functionality of the system under design. This source code must avoid architecture-dependent parts and can be written in any commonly used programming language such as C/C++, Java, SystemC, and others. Although this code could be written directly by the designer from the very beginning, the proposed methodology agrees with the need of applying software engineering techniques to speed up the process and to guarantee the quality of the resulting system.

One of the main aspects that should be addressed according to the proposed approach is the development of a high level system model, containing the specification of both the requirements and the solution itself. The requirements rely on the definition of three main aspects: (i) problem domain elements; (ii) desired behavior/functionality; and (iii) quality of service (QoS) requirements — performance, timing, power consumption, and cost. Once these elements are specified, designers can proceed with the development of the solution, which consists of a detailed description of the problem using the notation provided by the adopted modeling language and/or formalism, such as UML or FB.

Although the methodology itself does not rely on a specific language, this chapter provides some hints to help designers in deciding about which modeling language should be adopted. An important aspect to be considered is that the high level model should reflect the nature of the application domain. It is important and even necessary to use the most appropriate MoC for the model applicability to be enhanced. Furthermore, the language should be able to express both the application requirements and the functional specification. It should also provide facilities for allowing model validation prior to implementation, as well as features that can be used to guide the implementation. In order to clearly state such needs this chapter presents several qualitative and quantitative criteria for comparing the object oriented modeling approach of UML with the FB modeling approach provided by Simulink. A complete comparison is performed by means of a case study, which is presented in the next section.

Once the model is ready, mapping it to the architecture-independent specification is needed. In other words, it is necessary to generate the source code of the system under design. This process should be automatically carried out, but depending on the adopted modeling notation it may need different degrees of designer interaction. In the course of this chapter several existing CASE (Computer Aided Software Engineering) tools that can be used to help in this process will be discussed.

9.2.2 System Generation

The next step in the design process takes the architecture-independent description as input for an architectural exploration, where alternative hardware and software solutions which fulfill the system requirements should be considered and evaluated. After compiling all the available information the final system generation is performed, resulting in a micro architecture and a software description for a dedicated system.

In the SEEP methodology we are currently limited to accepting Java source code for representing the architecture-independent description. Using the SASHIMI environment [100] both a VHDL description for a dedicated Java processor and the respective program memory code (application code) are generated. This CAD environment automatically synthesizes an Application Specific Instruction Set Processor (ASIP) microprocessor for a target application, using only a subset of instructions used by the designed application. This Java processor implements an execution engine for Java in hardware through a stack machine compatible with the Java Virtual Machine (JVM) specification. Figure 9.2 presents the SASHIMI environment design flow.

9.3 Case Study: Evaluating High Level Models

This section presents two different models developed for comparing the object oriented modeling approach of UML with the FB modeling approach provided by Simulink. Our goal here is to analyze how suitable these two approaches are for the first phase of the SEEP methodology, namely, the architecture-independent design.

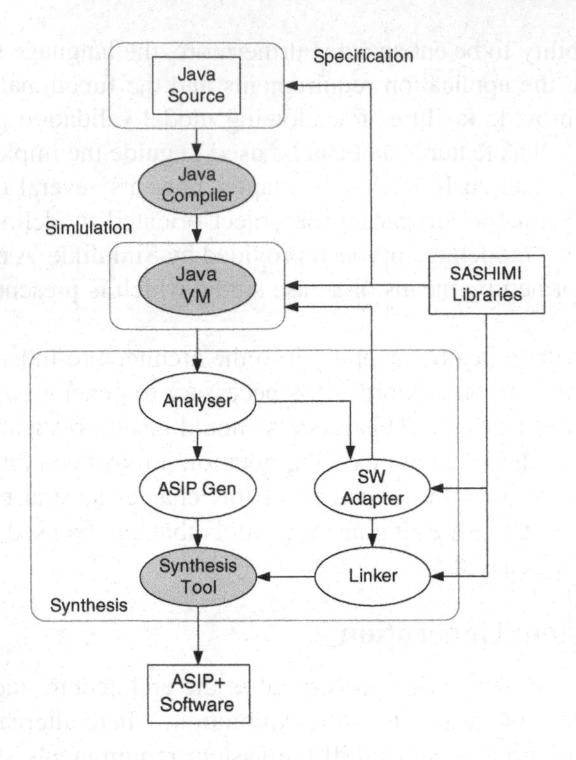

Figure 9.2. SASHIMI Environment Design Flow

The case study consists of a crane control system, proposed as a benchmark for system level modeling [137]. Once the user defines a position for the crane, the control system should activate the motor and move the crane to the desired point. Special care must be taken with speed and position limits while the crane is moving, in order to guarantee the safety of the transported load. Therefore

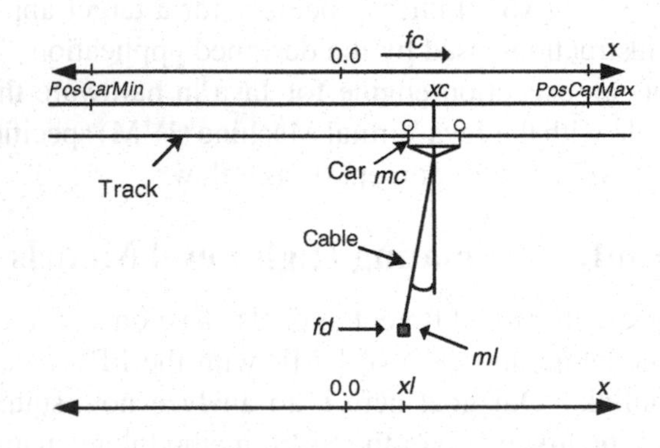

Figure 9.3. Crane System

constant monitoring is needed in order to avoid unexpected situations. This system incorporates hard real time constraints. Figure 9.3 gives an overview of the system.

9.3.1 FB Model

In the functional block (FB) approach applications are designed by connecting several FBs. Each FB output must be connected with an appropriate input, coming from a FB or another model element. This modeling language does not allow the designer to express system requirements. Therefore they start modeling already thinking of the solution for the problem under consideration. Our modeling starts with the functional decomposition, and the result is the definition of the modules that interact during the system execution. As shown in Figure 9.4, the modeling resulted in four high level modules organized hierarchically, as follows: PlantActuators; Sensors; ControlAlgorithm; and JobControl. Each module has its intrinsic behavior and is further detailed in this section.

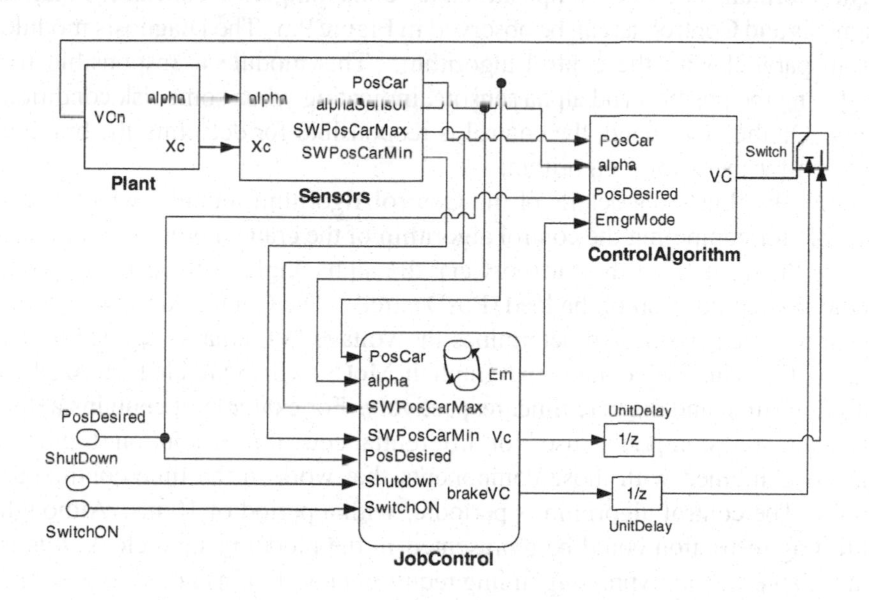

Figure 9.4. Crane Model using Simulink

The crane system is composed of both data driven and event driven parts, as can be observed in Figure 9.4. The JobControl module is represented by a finite state machine (event based), whilst the other modules are data driven. Figure 9.5 shows a view of the JobControl module that is composed by five states: Poweroff, Init, PosDesiredTest, NormalMode and EmergencyStop.

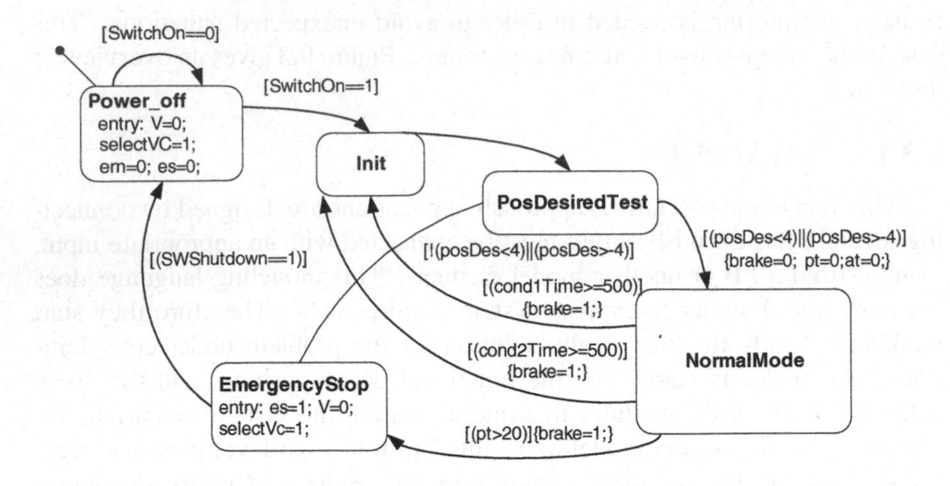

Figure 9.5. JobControl

The NormalMode is a composite state, containing two concurrent states, Diagnosis and Control, as can be observed in Figure 9.6. The Diagnosis module runs in parallel with the control algorithm. This module is responsible for monitoring the position and alpha sensors, indicating when some risk condition occurs. On the other hand, the control is responsible for detecting the braking condition for the control algorithm.

Figure 9.7 illustrates details of the ControlAlgorithm module, which is responsible for computing the control algorithm of the crane motor. This module receives the position of the car (posCar), the alpha angle of the cable (alpha), and the desired position of the load (PosDesired). The ControlAlgorithm computes a set of equations and determines the voltage (VC) that is applied to the crane motor. This FB contains two implicit MoCs, which are characterized as continuous time and discrete time, respectively. For example, it contains a discrete space state component used for differential equations resolution (top left), which is combined with those components that work in the time continuous domain. The control algorithm is periodic, with a period of 10 ms. Although this timing restriction could be represented in the model using a clock, this is not a suitable way of expressing timing requirements. For instance, no deadline can be stated, representing a missing piece of information required to perform schedulability analysis.

The Sensors module is responsible for reading the sensors and works with a fixed cycle time of 2 ms. Although this FB is not shown in this chapter, we observe that it has the same problems previously stated for the control algorithm regarding the representation of timing restrictions. Besides the position and angle sensors there are two other sensors for indicating when the car is beyond the track limits (minimal and maximum car position).

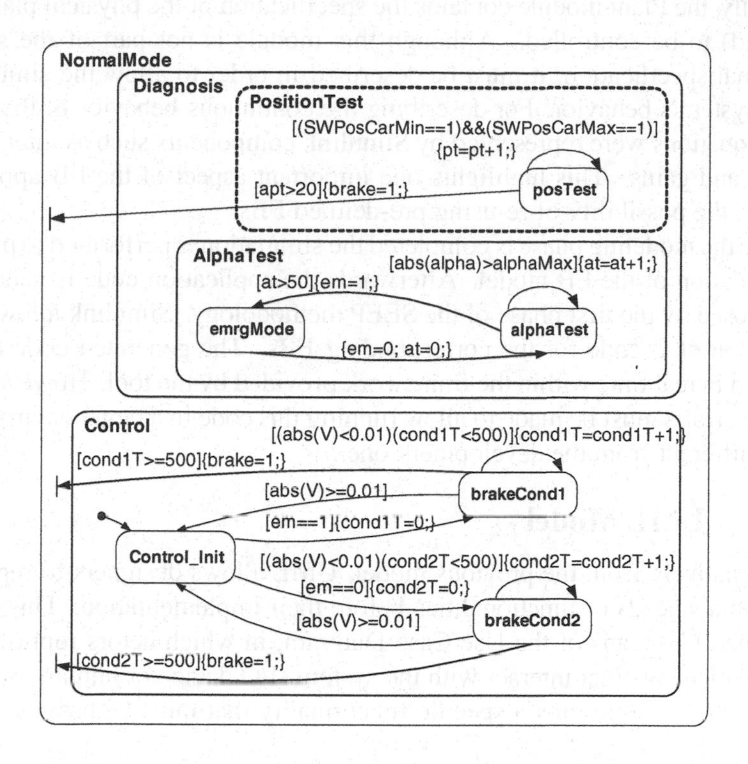

Figure 9.6. NormalMode

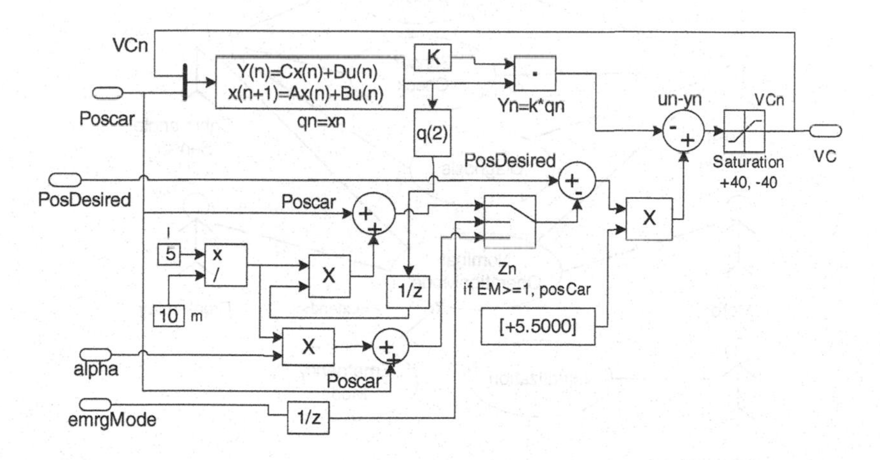

Figure 9.7. Control Algorithm Model in Simulink

Finally, the Plant module contains the specification of the physical plant (car and load) to be controlled. Although this module is not part of the system functional specification, it must be described in order to allow the simulation of the system's behavior. For describing the continuous behavior of the plant, linear equations were represented by Simulink components such as integrators, adders, and gains. This highlights one important aspect of the FB approach, which is the possibility of re-using pre-defined FBs.

Once the modeling phase is completed the simulation is performed to provide the validation of the FB model. Afterwards the application code is generated, as proposed by the first phase of the SEEP methodology. Simulink allowed the generation of C code for the corresponding FBs. The generated code can be executed in real time within the framework provided by the tool. However, reasonable efforts must be made to allow running this code in a target environment that is different from the development one.

9.3.2 UML Model

Alternatively from the previous model, UML allows designers to represent the system's needs or functionalities before their implementation. This can be performed by means of the Use Case Diagram, in which actors represent the external elements that interact with the system (I/O device or human user) and each Use Case represents a specific functionality that must be provided. The

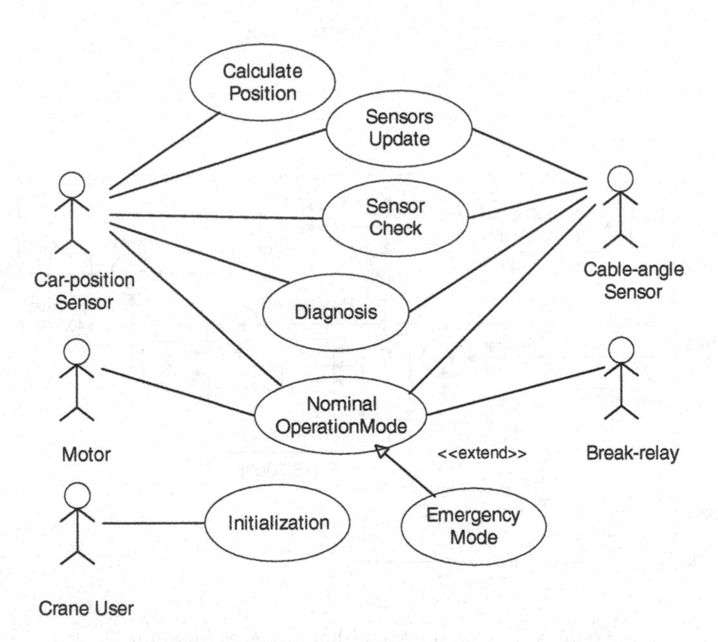

Figure 9.8. UML Use Case Diagram of the Crane system

Use Case Diagram for the crane system is presented in Figure 9.8. Each Use Case also includes a textual specification to detail its related responsibility. For a better structuring of the model development we followed the design phases proposed by Gomaa in the COMET/UML methodology [72]. However, any other UML based design methodology that considers real time aspects could be used.

To describe the interaction between objects that participate in each Use Case, they are further detailed using UML Collaboration Diagrams. This is part of the so called analysis modeling, which precedes the definition of requirements. To highlight important characteristics of the modeled system (mainly timing restrictions), the UML profile for Schedulability, Performance, and Time (SPT) [159] is used. This profile is also usually referred to RT–UML, and is composed mostly by stereotypes and its related tags. Using this profile, a timer event, for example, is decorated with the stereotype «SAtrigger». It includes information about its triggering frequency, as presented in the collaboration diagram from Figure 9.9 (see event 3 — run()). Such information is represented by the tag RTat of the stereotype that, in this case, means a periodic event with a 10 ms period.

Operations depicted in the diagram of Figure 9.9 represent the 'ControlAlgorithm' and, partially, the 'JobControl' blocks from the FB model (see Figure 9.7). Detailing the collaboration diagram one can see three different sequences of operations, denoted by the numbers 1, 2, and 3. Special attention is given to the third sequence, the control operation, which represents a periodic activity. Timing restrictions are denoted by the elements from the RT–UML profile. Similarly to the FB model, the controller class also has an associated state diagram, which is presented in Figure 9.10. This is part of the system dynamic model. One missing feature of UML noticed is the lack of semantics for allowing expressing the control algorithm itself, including its time continuous characteristics.

The complete UML model of the crane system includes 9 different collaboration diagrams. All classes from these diagrams constitute the system's static structure, which is used as input for the next development step from the COMET methodology, which is known as Design Modeling. This phase is responsible for defining the architecture of the system, including the division of responsibility between client and server objects. Since the crane model makes use of decentralized control, it was necessary to classify objects as being passive or active. The former represents data repository elements, whilst the latter represents elements with their own thread of control that are capable of triggering an interaction sequence. The final result is represented by the class diagram depicted in Figure 9.11. Classes names are preceded by '::' in order to follow UML conventions. They can also contain a stereotype incoming from the RT–UML profile (e.g., «SAschedRes», which denotes a concurrent element in the

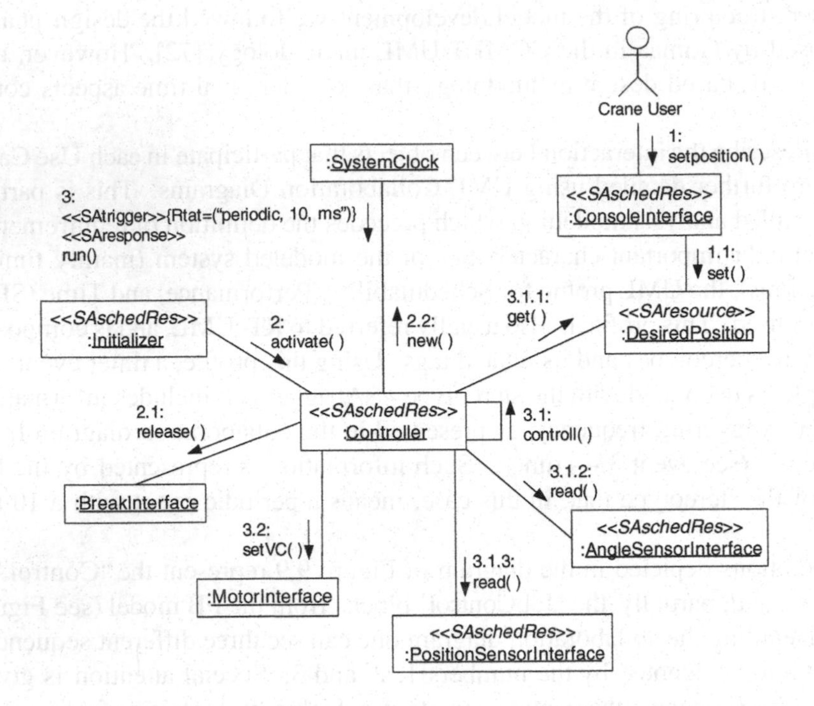

Figure 9.9. UML Collaboration Diagram of the Control Algorithm

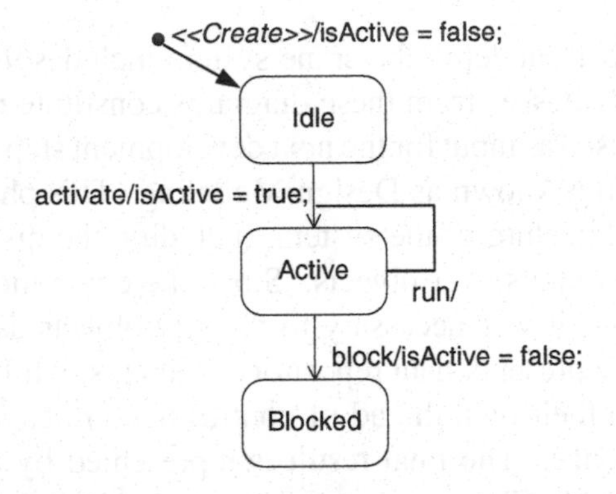

Figure 9.10. State Diagram of the Controller Class

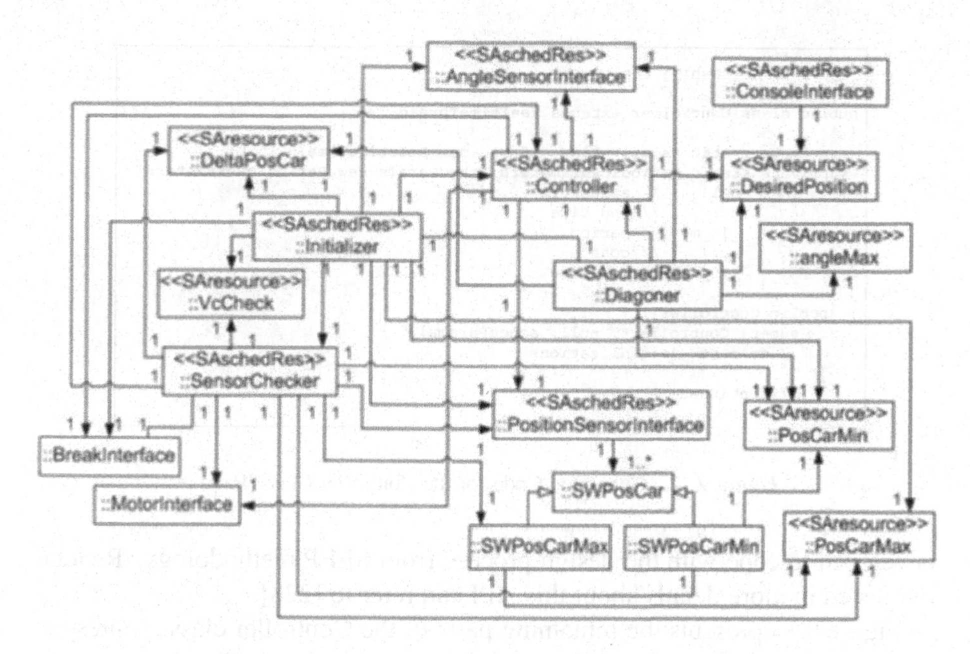

Figure 9.11. UML Class Diagram of the Crane System

system). The choice for the use of classes instead of capsules (part of UML 2.0) is due to the available runtime structure on which object communication is event based and does not use the port abstraction. This diagram is used as basis for the embedded system code generation.

As the design tool used to build the UML model did not include a simulation module, the next step was the code generation for the system. As expected by the SEEP methodology, an architecture-independent code should be provided. Although other programming languages such as C++ could also be used for code generation, the Java language was chosen as the target in this study owing to the current toolset used by the SEEP methodology (see Section 9.2.2). Details of the generated code will now be approached.

The Controller class, in which the associated stereotype denotes a concurrent real time task in the system, is selected to illustrate the generated code. This task is triggered periodically every 10 ms, with a deadline of 10 ms (see the collaboration diagram presented in Figure 9.9). To implement such features the Controller class needs to inherit features from RealtimeThread, as shown in Figure 9.12. Moreover, it must define release parameters to implement the modeled timing constraints. Therefore the PeriodicParameters class is used. Its instance is passed as a parameter for the superclass constructor. A RelativeTime object is used to represent the 10 milliseconds from the task period and deadline. All these special classes are derived from an API, which has been especially

```
import saito.sashimi.realtime.*;

public class Controller extends RealtimeThread
{
   private static RelativeTime _10_ms =new RelativeTime(0,10,0);
   private static PeriodicParameters schedParams=new PeriodicParameters(
            null,    // start time
            null,    // end time
            _10_ms,  // period
            null,    // cost
            _10_ms);// deadline

   public Controller() {
      super("Controller", null, schedParams);
      // do other initializations
   }
   ... //continues
};
```

Figure 9.12. Generated Code for the Controller Class (Part1)

developed to cope with the design process from SEEP methodology. Readers interested in more details about this API can refer to [228].

Figure 9.13 presents the remaining parts of the Controller class. It presents two important methods: mainTask() and exceptionTask(). The former represents the task body, that is, the code executed when the task is activated. Since this task is periodic, there must be a loop which denotes the periodic execution. The loop execution frequency is controlled by calling the waitForNextPeriod() operation. This operation uses the task release parameters to interact with the scheduler and control the correct execution of the operation. The exceptionTask() operation represents the exception handling code that is triggered in the

```
public class Controller extends RealtimeThread {
   ... //continuation
   public void mainTask() {
      Crane.breakInterface.release();
      // periodic loop
      while(isRunning == true){
         this.controll();
         Crane.monitorInterface.setVC(m_vc);
         this.waitForNextPeriod();
      }
   }

   private int controll() { ... }

   public void exceptionTask() {
      // handle deadline missing
   }
};
```

Figure 9.13. Generated Code for the Controller Class (Part2)

case of a deadline miss, that is, if the mainTask() operation does not finish until the established deadline.

After the code generation process the application was ported to the FemtoJava environment using the SASHIMI tool, and which was previously introduced in Section 9.2.2.

9.3.3 Evaluation Criteria

In order to perform a comparison between the modeling approaches, several evaluation criteria have been identified. These criteria are based on the work conducted by Ardis et al [4], which performs a qualitative comparison between several design languages for reactive systems. Such work is extended here in the direction of searching for aspects which could be used to perform a quantitative evaluation of the designed models. Moreover, a new organization for the set of criteria is established. They are organized into groups which reflect the needs observed in the first design phase of the methodology presented in Section 9.2, as can be seen in Table 9.1.

The groups are further refined into subgroups to compose the evaluation criteria elements. In Table 9.2 each evaluation criterion is detailed, together with an explanation of how it is evaluated (in qualitative or quantitative terms).

Table 9.1. Evaluation Criteria

Evaluation Criteria	**Description**
a)Requirements Specification	criterion for evaluating the capability of expressing and documenting user needs and system requirements
b) Functional Specification	criterion for evaluating the model abstraction level and expressiveness, i.e., if it describes the problem-domain elements and their behavior/functionality in a natural and straightforward manner
c) Validation /simulation	criterion for evaluating whether the specification can be validated before its implementation
d) Implementability	criterion for evaluating whether the specification can be easily refined or translated into an implementation, which is compatible with the rest of the system

9.3.4 Comparison Results

This section presents an analysis and comparison of the UML and FB models according to the criteria discussed in the previous section. The results are summarized in Table 9.3. For evaluating the qualitative aspects, we have used the symbol ++ to indicate a particular strength of the approach, + to indicate that the model meets the criterion in a way that is adequate, but less than ideal, and 0 to indicate a clear weakness of the model.

Table 9.2. Evaluation Criteria - Subgroups

Criterion	**Description**	**Evaluation**	**Expressed by**
a1) Functional requirements	capability of expressing and documenting the desired functionality of the system, together with the problem domain elements that interact with the system	Quantitative	the number of modeling diagrams that can be used to implement the desired feature
a2) QoS requirements	capability of expressing the application QoS requirements and/or restrictions	Quantitative	the number of QoS requirements that can be specified
b1) Applicability	capability of representing system behavior or functionality by using different MoCs, according to systems nature	Quantitative	number of supported MoCs
b2) Maintainability	easiness to make modifications in the specification, e.g., addition of new elements and changes in the external elements such as sensors	Qualitative	++ strength; + adequate; 0 weak
b3)Modularity / Hierarchy	capability of dividing a large specification into independent modules, which could be again decomposed into even smaller parts	Qualitative	++ strength; + adequate; 0 weak
b4) Expressiveness	capability of the modeling language primitives to describe the specification	Quantitative	b4.1) number of modeling primitives b4.2) number of different modeling primitives b4.3) number of lines of code programmed by the designer
c1) Simulation	capability of verifying if the specification can be used to validate the implementation	Qualitative	++ strength; + adequate; 0 weak
c2) Verification	capability of demonstrating formally that the specification or generated program meets the requirements	Qualitative	++ strength; + adequate; 0 weak
d1) Code generation	capability of generating an executable application from the model	Qualitative	++ strength; + adequate; 0 weak

Table 9.3. Comparison Results

Evaluation Criteria	FB	UML
a) Requirements Specification		
a1) Functional requirements	0	1
a2) QoS requirements	0	2
b) Functional Specification		
b1) Applicability	3	1
b2) Maintainability	+	++
b3) Modularity / Hierarchy	++	++
b4.1) Number of used modeling primitives	111	184
b4.2) Number of different modeling primitives	5	5
b4.3) Number of line codes written by the designer	0	96
c) Validation / Simulation		
c1) Simulation	++	+
c2) Verification	0	0
d) Implementability		
d1) code generation	++	+

This evaluation begins by analyzing the facilities for expressing the system's functional requirements. UML offers the facilities provided by the Use Case Diagram (1 point), in which functional requirements are defined in terms of actors and Use Cases. On the other hand, the FB approach does not support this kind of resource (0 points).

Regarding the support for QoS specification, one can see that the RT–UML profile supports both timing and performance requirements specification (2 points), whilst in the FB approach there is no support for such issues (0 points). In the FB model the timing requirements are implicit in the functional/behavior specification. Neither language give support to the specification of power consumption and cost requirements.

Analyzing the model applicability by means of the number of supported MoCs, it is possible to observe the advantages provided by the FB approach, as it supports three different MoCs (3 points): time continuous (analog), time discrete (digital), and event based. Regarding UML, it supports only the event-based model (1 point). In spite of this there are efforts described in literature that already address the lack of a dataflow model in UML (see [17], [76]).

Regarding maintainability, the intrinsic OO properties from UML models, such as the specialization/generalization facilities (inheritance), provide better maintainability if compared to the structured approach of FB models.

Considering modularity/hierarchy aspects, it is possible to conclude that the FB model leads to a slight better decomposition. This can be observed by comparing the Simulink high level model against the UML class diagram. The first one contains fewer elements, making the interpretation of the physical behavior easier. The UML class diagram used in our model maintains the whole system elements within the same abstraction level, which is somewhat unsuitable, considering the desired hierarchical features. However, the addition of the composite structure diagram in UML 2.0 overcomes this problem, since it allows for decomposition in a natural and straightforward manner.

The next criterion concern model expressiveness: number of used modeling primitives vs. number of different modeling primitives in use. The FB model contains 111 modeling primitives, except the plant module, including Simulink components, connections, ports, states, and transitions. In the UML model, 184 primitives are used. Regarding different modeling primitives in use, the UML model is represented by means of classes, objects, associations, states, and transitions. Therefore it is natural to observe an equivalent number of different modeling primitives if compared with the FB model, which includes blocks, ports, connections, states, and transitions. Nevertheless, using a design tool like Simulink, the designer can make use of different pre-defined components available in a component library.

Another relevant issue relates to the number of lines of code programmed by the designer in each model. It can be observed that in the UML model the designer has to manually write 96 code lines, whilst in FB model the program code was completely generated by the tool. Several UML tools have code generation capabilities, but they generate only code skeletons for classes and, at most, code from the Statecharts. The handwritten code parts include mainly the methods' behaviors that cannot be captured from the model. On the other hand, by using the FB model and associated library, the designer is not required to code the program by him/herself, as observed in our case study. Lastly, our experimental results show that by using a component library within the UML model it is possible to reduce the number of handwritten code from 96 to 66 lines.

Regarding model validation/simulation, it is possible to observe that in order to provide such features suitable modeling/design tools are required. Regarding the crane case study, only the FB model could be simulated, thanks to the Simulink tool which provides a simulation engine. The available version of the Real-Time Studio tool, used for the construction of the UML model, does not support model simulation. However, considering the authors' experience with other UML like modeling tools, they provide support at most for animation of Statecharts (event based MoC). Consequently, one can state that for this task the FB model is more adequate, because the simulation environment supports all the three intrinsic MoCs.

Analyzing the verification features, neither UML nor FB approach have support of formal verification of complete models. In UML some tools allow for model checking in specific diagrams, such as Statecharts and Sequence Diagrams. Moreover, many tools support consistency checking between diagrams, for instance checking the connections between the components in a FB diagram or even guaranteeing that an operation called in a UML collaboration diagram exists in the related class. For this reason both languages are considered weak in this aspect. Moreover, UML commercial tools check the syntax of actions in the Statecharts or if an operation called in a collaboration diagram was defined in the class. Therefore Damm and Harel [46] proposed the Live Sequence Charts (LSC) that are a extension of Message Sequence Charts (MSC) with rigorous semantic. The use of the LSCs allows consistency check between the generated scenarios and the sequence charts applying formal verification techniques.

Finally, considering the model's implementability one can see that from both models an architecture-independent specification can be derived. Still, there are two aspects that lead to distinct capabilities: amount of code provided by designer and number of pre-defined components. In UML the need for designer intervention is higher as can be observed in the crane case study, because some parts of the specification cannot be expressed using UML diagrams (e.g., the control algorithm). In the FB models the whole code can be generated automatically, since it relies on the use of pre-defined libraries. However, the generated code requires several modifications/optimizations to be executed outside the framework provided by Simulink.

9.4 Problems in Mapping from High Level Models to HW–SW SoC

The SEEP methodology proposes an approach for the synthesis of a dedicated SoC that starts from a high level model model, as illustrated in Figure 9.1. This approach suggests the mapping of the high level model to a Java application, which is further synthesized in order to obtain a dedicated SoC comprising software and hardware components. This section discusses problems related to the transformations of FB and UML models into a HW–SW SoC according to the SEEP methodology. The following is illustrated in Figure 9.14.

The main problem observed in the proposed mapping is the so called 'MoC semantic gap', i.e., the lack of abstractions in the modeling language for representing properly a certain feature of the system. Another problem regards the 'generality' of the specification, i.e., how independent from architecture and platform it is. Both problems are detailed in the next subsections. Moreover, this section summarizes the results from the synthesis of the UML model of the crane system.

9.4.1 The MoC Semantic Gap

This problem relates to the lack of abstractions in the modeling language for representing properly a certain feature of the system. In our study this gap is more evident in UML since it does not provide a natural abstraction for modeling

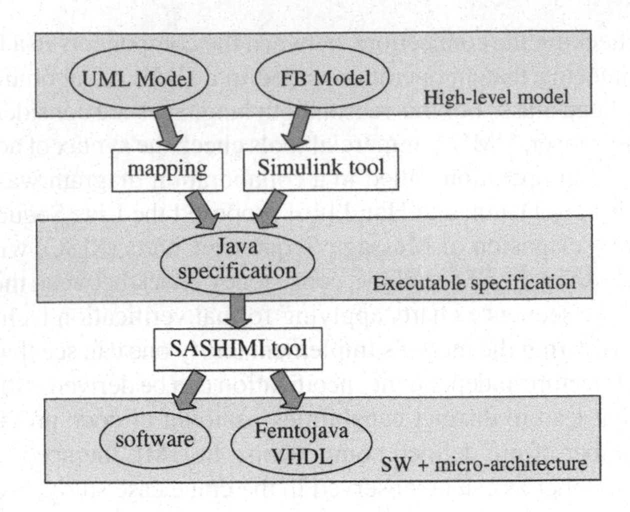

Figure 9.14. Proposed Design Flow

some specific behaviors, such as the control algorithm. The main reason for this problem is that UML is only event based and consequently not capable of representing the continuous and discrete time MoCs, as required by the control algorithm. Therefore the designer needs to write the related algorithm using an adequate programming language, thus reducing the model abstraction. The consequence is that the resulting model will not be 'pure' UML, but UML skeletons filled with handwritten code. Related approaches that tackle this gap are presented in Section 9.5.

9.4.2 The Model Generality Problem

It is possible to observe another problem which, at least in UML, is directly related to the MoC semantic gap. During the design phase the model is filled with code written in a specific programming language. This makes the model strongly connected to this language, i.e., the model is composed by UML plus the programming language (C++, Java, etc). The result of this is that it is not possible to make code generation for a different target language. For that reason one needs to develop a new model, rewriting the code within the UML model. The imminent conclusion is that UML should have its own semantic for expressing the application's behavior to allow the design of complete platform-independent models.

Therefore there is a growing trend towards building systems to different target platforms directly from models. To enable rapid prototyping it is necessary to have models designed in a language-independent manner. In this context the OMG promoted the definition of UML action semantics [148], which, when combined with UML, allows modelers to define behavioral specifications at a higher level of abstraction. The main advantage is that the UML action semantics is independent of any specific underlying technology in the execution environment. The adoption of the precise action semantics for UML supports the viability of executable UML (xUML). Unfortunately, the precise action semantics for UML is a semantic standard only, not a syntactic standard [25]. It is easy to see that competing syntaxes may constrain the ability of xUML to gain a mind share in the development community and may, by extension, hinder the adoption of this technology. Given that, alternative solutions have been proposed. An example is the Object Action Language (OAL) designed by Mentor Graphics [131], which is platform-independent and allows the specification of actions associated to all states of the Statecharts. Other related approaches are presented in Section 9.5.

From the FB model the whole code can be generated almost automatically, since it relies on the use of pre-defined libraries and details of the implementation have already been defined in the model. Nevertheless, the FB model also has limitations related to the possible target platforms. Parts of the models, such as the control state machines, can be mapped to different target languages, but this does not apply to the all FBs, which are mostly language-dependent. Additionally, if the Simulink tool is used for modeling the system, an executable specification can be automatically derived thanks to this tool having a library of pre-defined components. However, this executable code cannot be cross-compiled to the embedded platform, since the pre-defined components are suitable only for the host platform and its provided framework.

9.4.3 Results Summary from the Crane Synthesis

In this work we have focused in the path from a high level model to an SOC using the SASHIMI tool in order to synthesize a dedicated processor. This tool generates a VHDL description of a Java processor (femtoJava) and the application software for this processor. Because the Simulink tool was unable to generate the Java code of the model, the obtained results reflect only the synthesis from the UML model.

Table 9.4. Synthesis Results

Area (μp size)	3749 LCs
RAM Size	780 B
ROM Size	8 KB

Table 9.4 presents the results of concrete experiments in terms of memory size (program memory and data memory) and processor size. The area taken by the processor was computed in number of FPGA logic cells, after synthesis from the VHDL description of this processor. These results were obtained for an FPGA synthesis, but the VHDL description could be also used for the

generation of an ASIC version. The approximate area in gates of this processor is around 50k gates. The results were obtained using Xilinx ISE version 6.2i [232] and the FPGA used was the Xilinx SpartanIIE XC2S200E-6PQ208.

9.4.4 Modeling Approach in the SEEP Methodology

After considering the results of the performed analysis we chose to use UML as a modeling language in the SEEP methodology for the following reasons: (i) UML is a recognized standard and is becoming very popular, especially in industry; (ii) UML offers extension mechanisms (stereotypes, tagged values, constraints) allowing us to implement our own additional semantics and adding it into UML; (iii) UML is modeled by a metamodel, itself specified by the MOF [149]. This allows us to specify a metamodel that extends UML. The conformity of models exchanged between tools is ensured by the XMI/XML standard [155]. Using OO concepts of UML a definition of a class is made of a class interface and a class behavior. This distinction between definition and instances allows the development of libraries of re-usable components. Another contribution from OO is the ability to define a component by inheriting features from another one, which again improves the re-use of components.

Moreover, UML does not depend on any particular methodology. For this reason we can define our own methodology. Other projects in the area of real time systems or embedded applications have also chosen UML as the modeling language. For example, the HASOC methodology [75] extends UML-RT to include annotations with mapping information. In this work, the authors propose the association of capsules with additional MoCs, such as Synchronous Dataflow and Codesign Finite State Machines. Another research group proposed an UML profile for embedded system platforms [36], which allows the modeling of platforms, quantifying QoS performance and budgeting constraints, and revealing platform services. In this work an IIR filter was modeled using the proposed profile to demonstrate its use in applications targeted to the wireless domain and other domains with intensive use of digital signal processing techniques. The model strategy used in the IIR example is difficult to understand, mainly because it is hard to see a direct correspondence between the filter UML model and its describing equation. Moreover, the model is too verbose, since it uses several modules to describe a simple equation with two multiplications. Additionally, the model abstraction level is very low, going down to the micro-operation level, and is not adequate for complex embedded system modeling.

9.5 Related Works

At present, the authors are not aware of similar work comparing the OO modeling approach from UML against the FB modeling approach. However, there are several proposals for combining both modeling paradigms. In [91] a profile for integrating FBs into UML is proposed. Therefore the General Function Block Model is presented, working as a kind of adapter between classes and FBs. Another work [17] addresses the lack of a dataflow model in UML and presents a proposal for the integration of both modeling paradigms. More recently Green et al in [76] included the support of dataflow model into activities UML diagrams, but this work is recent and is not completely validated.

Additionally, other works observe that UML is not suitable for representing other MoCs besides the event-based one. For that reason other extensions are proposed. Axelsson [8] proposes an UML extension to represent continuous time relationships, such as continuous variables, equations, time, and derivatives. A similar work has been conducted by Berkenkötter et al [15], in which the authors extended UML using as its basis a programming language suitable for hybrid systems.

Recently a new visual language called Systems Modeling Language (SysML) was submitted to the OMG [207]. It re-uses a subset of UML 2.0 and extends the language so as to satisfy the requirements of the UML for Systems Engineering (SE) domain. In this proposal the activity diagram is extended to support traditional SE functional flow block diagrams (dataflow) and continuous behaviors.

In order to support model verification, simulation, and synthesis, Bjorklund [20] proposes the use of Rialto as the intermediate language during the model's design. This language has a formal semantics that allows the capture of the semantics of UML behavioral diagrams. As a result the language can be used as an execution engine for UML models and also to generate code. Rialto can also be used to combine multiple MoCs because different scheduling policies are defined in this language. In this work the authors consider that the activities diagrams have dataflow as their underlying model of computation and these diagrams can be interpreted as a Statechart in which all computation is performed in state activities and the transitions are triggered by completion events. However, a Statechart is control-flow like and is not an adequate representation for dataflow models. Moreover, as this is an ongoing work, it supports only some UML diagrams.

Hubbers and Oostdijk [95] highlight the gap between a model and its implementation, pointing to the difficulty of verifying whether the implementation's behavior agrees with the specification. In this context the authors propose the use of JML (Java Modeling Language) specifications in order to facilitate this verification. A JML specification allows one to formally verify whether the generated code implements the specified model. Tools which are compatible with this language are available, such as ESC/Java [62] and Loop [101]. In this project a tool called AutoJML has been developed, which automatically derives JML specifications from UML state diagrams represented in the XMI format, beyond the Java code. The combination of the JML specification and the skeleton code can be formally verified using the ESC/Java.

In [170] one can find a proposal for the integration of different modeling approaches using UML metamodels, allowing that the designer chooses the most adequate tool to model a given part of the system. Models developed with different tools can be integrated, and an executable code can be generated from this integrated model. In this approach an action language called MeDeLa is used to specify the operation behavior. This proposal is compliant with the action semantics proposed by OMG. The code generator generates Java or C++.

9.6 Conclusions

This work has presented a comparison between FB and UML for high level modeling according to the SEEP methodology. Considering the result obtained, it can be seen that UML works better for requirements specification. Nevertheless, none of the models deal properly with the specification of embedded systems requirements, since power, for instance, is not included. An advantage of UML is that it can be extended to incorporate this and other features. Comparing the developed functional specifications, a similar score is observed for both approaches. This leads to the conclusion that models are somehow equivalent, although each has pros and cons in this aspect.

Although OMG promotes efforts towards the precise action semantics, such as xUML, it does not have a syntactic standard. This weak aspect results in the use of different syntaxes in the UML based CAD tools. In the course of this chapter several options for representing the system behavior were presented. Our approach, for instance, uses Java code within the model to describe the algorithmic behavior. From the generated Java code one can synthesize a dedicated Java processor. Results obtained for the ASIP generation from the UML model, containing processor and memory size, have been presented.

Lastly, our conclusion is that UML is more suitable for being used within the SEEP methodology. This is owing to the following reasons: (i) UML is a recognized standard and is becoming very popular, especially in industry; (ii) UML offers extension mechanisms (stereotypes, tagged values, profile) allowing us to implement our own additional semantics and to add it into UML; (iii) UML is modeled by a metamodel, so that the conformity of models exchanged between tools is ensured. Moreover, UML does not depend on any particular methodology, making it suitable for being used in our approach. Additionally, other projects in the area of real time systems or embedded applications have also chosen UML as their modeling language. Moreover, it provides all benefits from the object oriented paradigm such as modularity, encapsulation, re-usability, concurrency, etc.. Therefore its use should be widely expanded in the new generation of SoCs and embedded systems.

Chapter 10

A Model-Driven Development Process for Low Power SoC Using UML

Yves Vanderperren, Wim Dehaene

Department Elektrotechniek – ESAT – MICAS
Katholieke Universiteit Leuven
Leuven, Belgium

Abstract UML is gaining increased attention as a system design language. This is confirmed by several reported experiences and current standardization activities such as the SysML initiative and the UML for SoC Forum in Japan. The adoption of UML 2 is a significant step towards a broader range of modeling capabilities. This chapter analyzes the impact of these recent advances on the applicability of UML in SoC development. An application of such new practice is presented in a model-driven development process geared towards SoC design and taking benefit of the best techniques recently introduced. In addition, the crucial issue of power efficient system design with UML 2 is investigated.

10.1 Introduction

Larger scale designs, increased mask and design costs, 'first time right' requirements and shorter product development cycles motivate the application of innovative 'system on a chip' (SoC) methodologies which tackle complex system design issues. There is a noticeable need for design flows towards implementation starting from higher level modeling. However, the concepts of system level modeling, analysis, and refinement are subject to various interpretations, owing to the wider scope of application at higher abstraction levels. Several entry points into the flow can be defined from high level modeling languages and abstraction levels.

UML for SoC is attracting growing interest in the recent years, in particular since 2002 [125, 74, 47, 139, 220], and the conclusions from different experiences in the industry have already been published [164, 64]. The fusion of some of the best ideas from the digital hardware and software engineering domain provides major benefits. A common and structured environment for capturing

G. Martin and W. Müller (eds.), UML for SOC Design, 223–252.

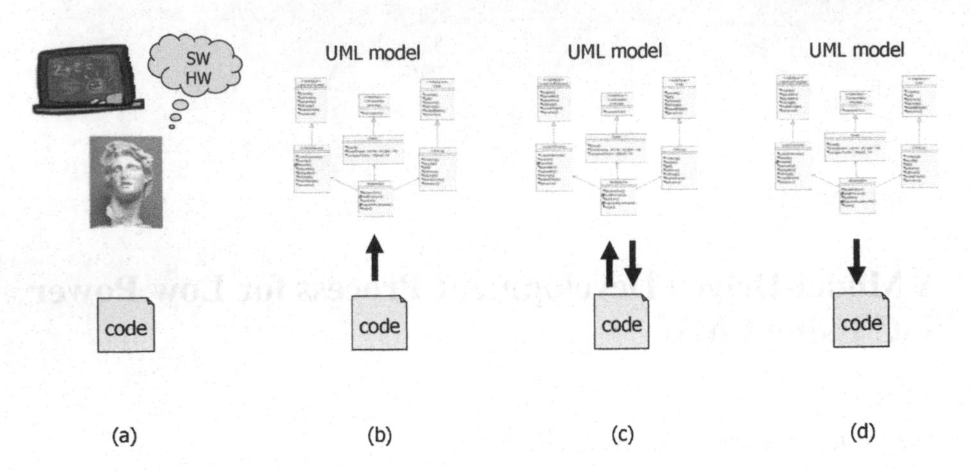

Figure 10.1. Relationship between Model and Code: (a) Code Only; (b) Code Visualization; (c) Round Trip Capability; (d) Model-Centric Approach

the requirements, a unified view of the system, a notation complementary to SystemC are some of the major advantages identified while using UML.

Significant issues remain, however. The most important is the need to customize UML towards the specificity of SoC development. Following the trend of platform-based designs, a UML profile to model embedded system platforms has been proposed [36]. Even so, development tools are slow to realize UML's full potential in a SoC context, despite the success of UML in its main areas. More recently the UML for SoC Forum [88] in Japan has been working on a specific extension in the form of a UML profile for SoC to solve these shortcomings and improve interoperability of tools.

Furthermore, UML requires the definition of model elements with clear semantics geared towards SoC design languages, in order to efficiently support model-centric design flows for SoCs. Several works have already proposed extensions to UML towards SystemC [12, 143, 173]. The first benefit is *code visualization* through graphical notation to aid understanding of the code structure and behavior (Figure 10.1). As a next step *round trip* capability between the code and the UML model becomes feasible if these customizations are standardized and supported by existing tools. The system models contain enough information in *model-centric* approaches to allow complete code generation. This situation follows the principles promoted by the Model Driven Architecture (MDA) initiative of the Object Management Group (OMG) [132] and the concept of executable and translatable UML (xtUML) [131, 168]. MDA encourages the development of platform-independent (PIM) UML models which are mapped onto platform-specific (PSM) models. This approach can be seen as a particular case of model-driven development that emphasizes the role of (executable) models. However, SoCs can present severe real-time and power constraints which limit the applicability of MDA and the generation of the system implementation. Moreover, deep sub-micron effects threaten classical development approaches in which design is separated as much as possible from manufacturing aspects. As a result the principle of decoupling functionality from implementation becomes even more problematic with nanometer technologies and the increasing intrusion of device physics into design [39].

Finally, employing a modeling language by itself is not sufficient. Questions such as the pros and cons of UML, the available tools and their interoperability are frequently addressed. But the impact of a well defined development process which assists engineers in improving their design using UML is often neglected. UML provides a notation for modelling systems, allows one to couple the model with conceptual and executable artifacts, addresses complex systems with scalable and flexible modeling means, and improves the communication between stakeholders. UML is neither a methodology nor does it dictate any particular development process[1] to be used, but provides communication support within the development process, which is the backbone of the project.

This chapter stresses, therefore, the importance of a sound methodology guiding the system design flow from the requirements capturing phase to the start of detailed hardware and software implementation, assisted by the modeling capabilities of UML 2 and SysML, an extension of UML towards systems engineering. Their essential features are summarized next before considering a concrete case study.

10.2 UML and System Design

10.2.1 From UML 1.x to 2

Several deficiencies were identified when applying UML at system level, as reported, e.g., in [70]. UML 2 alleviates some of these issues. The UML 2 specification is in its finalization phase [154] and represents a significant milestone in the evolution of UML to support the design of complex systems. Some concepts from UML-RT/ROOM [190] are now standard, such as the notion of ports. The introduction of the composite structure diagrams eases the representation of the structural aspects of a system. The improved support for hierarchical decomposition of the system structure facilitates the representation of systems which are not necessarily object-oriented in their nature. The capability to model complex and parallel behavior has significantly improved with UML 2, which supports the hierarchical decomposition of the behavioral constructs. Sequence diagrams can have fragments and allow the representation of loops, alternative sequences, and parallel message exchanges. Activity

[1] A process describes how to organize work in a common direction and provides a structured set of steps to design successfully robust systems which efficiently meet the customer requirements.

diagrams are now based on Petri nets and the representation of concurrent flows of operation is improved. Timing diagrams stress the importance of time when showing the interaction between objects and their change in state.

Although UML 2 represents a significant step towards system modeling, it still presents several imperfections. One of the most significant problems is the question of the semantics of UML.

10.2.2 UML and the Issue of Semantics

UML is the result of a converging trend from the 'method wars' towards a unified notation. As a result a common criticism of UML is the wealth of available diagrams. Instead of simplifying this situation UML 2 adds four more diagrams to the nine existing ones in the previous version of UML.

A more severe concern with UML is its lack of semantics[2]. Although UML standardizes the syntax of the diagrams it does not provide the detailed semantics of the implementation of their functionality and structure. Precise semantics is required to ensure efficient communication, to provide a notion of well formed model, and to enable automated code generation. On the other hand, providing too detailed precision would limit the applicability of UML. The domains in which UML is now applied are so different that they cannot be unified under a single semantics. UML therefore remains a semi-formal language and allows *semantic variation points*, i.e., its semantics may vary according to the application area.

Two opposite directions of solutions can be identified. A first approach consists in defining precise semantics for a subset of UML based mainly on class diagrams and simplified state machines. Since UML 1.5 and the adoption of a precise action semantics the behavioral aspects of UML models can be specified with enough details to allow an executable application to be generated from the UML model. The semantically defined set of operations involve direct manipulation of UML modeling elements and are specified by means of a text based *action language*. Owing to the lack of standardized syntax, action languages vary between tool vendors. Nevertheless, using an action language allows one to specify the behavior of a system at a very high level of abstraction and is independent of any execution environment. Model compilers translate the platform-independent model by mapping it onto a target implementation. This approach is close to the MDA philosophy and gives rise to the different flavors of executable and translatable UML (xtUML).

Instead of considering a subset of UML with precise semantics, another point of view is to provide domain-specific semantics in dedicated UML profiles. A

[2]The way symbols look like and can be combined forms the *syntax* of a language. The exact meaning and interpretation of each symbol constitute the *semantics* of the language. More details can be found e.g. in [87].

profile is a UML package which defines sets of extensions for a particular domain or purpose. It allows one to create new modeling constructs based on the existing ones within UML. As a result a profile does not create a new modeling language. This would be the case if a new metamodel is defined from the Meta-Object Facility (MOF) in a similar way to that in which UML is specified. The motivation behind the use of a profile is the need for a model to capture accurately the essence of the considered domain. Since UML is conceived as a semi-formal language and a general purpose notation, modelers may have to bring domain specific adaptation to the language in order to define particular semantic elements of interest and to fit their needs. However, profiles present the risk of transforming UML into a modeling paradigm instead of a language. For instance, several different profiles have been proposed to bridge the gap between UML and SystemC [12, 143, 173]. Having a standard extension would benefit the SystemC community by providing common communication means. On the other hand, no profile is likely to suit all the needs of each user. In this aspect a sound and suitable development process for guiding designers in the application of modeling languages, in particular UML, plays a major role.

Several profiles are already ratified by OMG, such as the UML Profile for Schedulability, Performance and Time[3] [159], or are under standardization procedure, e.g., the UML for SoC profile or SysML [207]. The latter is introduced in the following section and applied later to a concrete case study.

10.2.3 SysML

Motivation. The Systems Modeling Language (SysML) is a joint initiative of OMG and the International Council on Systems Engineering (INCOSE). The purpose is to refine UML and provide a general purpose modeling language for systems engineering. This field covers complex systems which include a broad range of heterogeneous domains[4], in particular, hardware and software.

Strong similarities exist between the methods used in the area of systems engineering and complex SoC design, such as the need for accurate requirements capture, heterogeneous system specification and simulation, system validation and verification. Several features from SysML can be applied to SoC design.

Overview of SysML. The SysML initiative will result in a profile extending UML 2 to systems which are not purely software based. Although UML is

[3]Also called the Real-Time Profile (RTP).

[4]INCOSE [98] defines systems engineering as an "interdisciplinary approach and means of enabling the realization of successful systems. It focuses on defining customer needs and required functionality early in the development cycle, documenting requirements, then proceeding with design synthesis and system validation while considering the complete problem. Systems engineering integrates all the disciplines and speciality groups into a team effort forming a structured development process which proceeds from concept to production to operation.".

currently broadly defined[5], many concepts remain tightly linked to software applications and are more relevant from a software perspective, such as deployment diagrams. Whilst UML 2 provides stronger support for modelling hierarchical system architectures, for example by means of composite structure diagrams, an ambiguity remains regarding the choice between component [45], composite structure and deployment diagrams [74, 236], when considering systems like SoCs in which the description of the hardware part may be as important as the software part. Moreover, flows such as a discrete stream of data cannot be represented on any of these diagrams. SysML proposes a unified and domain-neutral solution to represent the system architecture and flows of information by means of SysML *assembly*, which is a stereotyped class describing a system as a collection of parts with specific roles. The SysML assembly can be used to represent both the logical and the physical aspects of various of kinds of systems. Since an assembly is domain-neutral, it can be applied to describe the hardware as well as the software part in the SoC context. An assembly can represent a black box view of a system without showing the internal structure, as well as a white box view in which the internal parts, connectors, and ports are visible. SysML considers the 'ball and socket' notation for provided and required interfaces as being specific to the software domain and is therefore not used by an assembly.

In addition SysML provides support for representing requirements and relating them to the model of a system, the actual design and the test procedures. Although tracing the requirements of a system from informal specifications down to the individual design elements and test cases is a fundamental activity, UML does not address this subject in detail. A Use Case (UC) analysis typically helps build up a sound understanding of the expected behavior of the system and validate the proposed solution. However, requirements are often only traced to the use cases but not to the design. Consequently the focus on the responsibility of the different parts of the system is lost and the verification becomes more complicated. SysML therefore introduces the *requirement diagram* and defines several kinds of relationships for improving the requirement traceability. The purpose is not to replace the numerous commercial tools dedicated to this subject, but to provide a standard way of linking the requirements to the design. Requirements can be decomposed by means of the *containment* relationship in a similar way to that for class diagrams. The *trace* dependency relates derived requirements to source requirements. The system designed and the requirements are linked by a *satisfaction* dependency. Finally, the *verification* dependency associates a requirement with the test case used to verify

[5]"The Unified Modeling Language is a visual language for specifying, constructing, and documenting the artifacts of systems." [157]

this requirement. Examples of stereotypes defined for these relationships are shown in the case study later in this chapter.

SysML introduces one more diagram, the *parametric diagram*, to model properties and their relationships such as mathematical expressions or constraints. This diagram provides more benefits in the scope of mechanical engineering than SoCs.

Although SysML adds two diagrams to UML, it abandons the use of the communication diagram and the deployment diagram. The latter is considered specific to the software domain and limited to the deployment of software artifacts onto hardware components. The SysML version of the composite structure diagram with the support for flows is meant to provide an equivalent and domain-neutral modeling capability. Furthermore, the concept of *allocation* in SysML is a more abstract form of deployment and is defined as a design time relationship between model elements which maps a source onto a target. An allocation can associate miscellaneous elements. For example, the link between the requirements and the design elements (the *trace* and *satisfaction* dependencies) is a *requirement allocation*. An allocation also applies to the mapping of a function onto the structure implementing it (*functional allocation*) and logical assemblies onto physical assemblies (*behavioral allocation*).

Finally SysML provides several enhancements to activity diagrams. In particular the control of execution is extended such that running actions can be disabled. In UML 2 the control is limited to the determination of the moment when actions start. In SysML a behavior may not stop itself. Instead it can run until it is terminated externally. For this purpose SysML introduces *control operators*, i.e., behaviors which produce an output controlling the execution of other actions.

More information regarding the detailed features available with SysML can be found in the SysML specification [207].

Shortcomings of SysML. SysML contributes to more rigorous transfer of the specifications between system, software and hardware engineers, introduces several modeling elements applicable across multiple domains, and enhances UML 2 toward systems engineering. Despite these benefits SysML falls short in the following points.

Use cases are not well integrated with the rest of UML [104]. Although SysML stresses the traceability from the requirements to the design, it does not help in solving this issue and the difficulties associated with use cases [70, 120].

SysML provides domain-neutral modeling capability which allows a high level understanding of systems. However, these modeling constructs are used at the level of system conceptualization and requirement engineering. They must be linked later during detailed design to domain-specific model elements—for example, the UML representation of a SystemC model of the system, or the

UML model of the software architecture designed. It remains unclear how the concept of *allocation* can cope in this situation with notations excluded by SysML, such as the 'ball and socket' notation which is applicable to software. Moreover, SysML does not provide domain-specific semantics and does not completely solve the semantics issue raised above.

SysML is recent and still under evolution. More lessons from practical experiences are needed and the standardization effort of SysML would benefit from an involvement from system houses (from the telecom area) and semiconductor design companies, which could complement the current contributions from the space, aircraft, and defense industry.

SysML is only a modeling language like UML and does not provide any development process. SIMILAR [9] is an example of parallel and iterative process based on a study between several approaches in the area of systems engineering. Processes are discussed further in the following section.

10.3 Process and Model-Driven Development

UML defines a notation but does not include any development process or life cycle model. A well defined process provides significant advantages. First, it coordinates the activity of the project team. This ensures a consistent evolution of work. Second, a process eases the measure of progress by identifying the deliverables at crucial steps and provides effective means of planning. Third, it encourages continuous assessment of the product designed and the process itself. Modern development processes for software [106], embedded software [49], and systems engineering [9] follow Boehm's spiral model [22] and are structured around a number of iterations or microcycles. Each of these involve several disciplines of system development running in parallel, such as requirements capturing, analysis, implementation and test. The effort spent in each of these parallel tasks depends on the particular iteration and the risk to be mitigated by that iteration. As a result large scale systems are incrementally constructed as a series of smaller deliverables of increasing completeness which are evaluated in order to produce inputs to the next iteration. This continuous feedback loop improves not only the quality of the final product but the process as well, which is nothing else than a development model.

It is crucial to adapt the development process to the particular needs of the given application domain. In the SoC context, characterized by severe first time right requirements and exponentially increasing mask costs which prevent successive physical implementations, executable models provide a means to support iterative development process. This specificity makes SoC development processes like [219, 236, 74] resemble systems engineering processes, such as SIMILAR [9] (Figure 10.2) or HARMONY [94]. Abstract and executable models play a crucial role in these development processes. First, they resolve

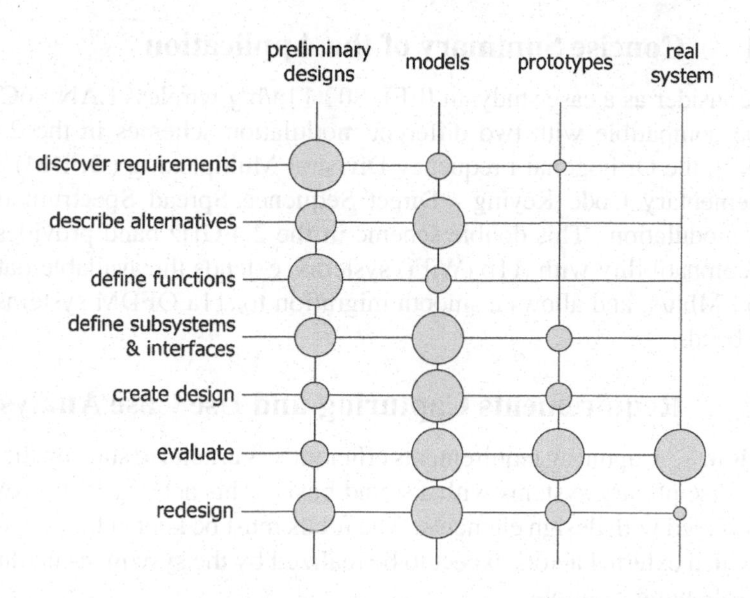

Figure 10.2. Amount of Emphasis in the Phases of the Development Lifecycle

ambiguities related to paper specification. Next, they capture the essential characteristics of the large and complex system for design, and allow designers to concentrate on the key issues. Third, executable models help analyze the system and spot performance bottlenecks of the architecture. The bottom line is that models provide a simplified representation of a system in order to validate and verify performance in early stages of the development. During *validation* the system is evaluated against the specified requirements to guarantee that the right system has been built. The *verification* task ensures that the system is built correctly, it focuses on systematic errors, deadlock conditions, etc., which must be absent in the designed system. The *performance* is the degree to which the system accomplishes its designated functions within given constraints such as speed and power. Models help identify performance issues but can never absolutely prove the absence of problems, since a model is a simplification of reality. Models must capture the essentials characteristics of the designed system. Executable models rely on a language having precise semantics and play an essential role in avoiding ambiguity and self-contradictory models.

10.4 Model-Driven Development for SoC: Case Study

This section illustrates by means of a concrete case study the application of a SoC development process using UML. Although such a process spans the whole design until completion of the project, this chapter focuses on system level design and the development of an executable model of the architecture until the start of the implementation of the detailed hardware and software.

10.4.1 Concise Summary of the Application

We consider as a case study an IEEE 802.11a/b/g wireless LAN SoC which must be compatible with two different modulation schemes in the 2.4 GHz band [97], the Orthogonal Frequency Division Multiplexing (OFDM) and the Complementary Code Keying / Direct Sequence Spread Spectrum (CCK / DSSS) modulation. This double scheme in the 2.4 GHz band provides backwards compatibility with .11b (WiFi) systems, extends the available data rates up to 54 Mbit/s, and allows a smooth migration to .11a OFDM systems in the 5 GHz band.

10.4.2 Requirements Capturing and Use Case Analysis

Overview. Capturing requirements efficiently is crucial to starting the design of multidisciplinary systems with a sound basis. This activity is, however, too often mingled with design elements. The focus must be kept at this stage on the goals which external actors expect to be realized by the system, as is illustrated by the following example.

Figures 10.3, 10.4, and 10.5 show a typical architecture of a complete 802.11 a/b/g solution. We assume the MAC to be an available IP to which the PHY layer's capability is appended, including both hardware and software parts. The MAC is supposed to have a general purpose processor which may be shared with the PHY, as shown by the dashed lines in 10.5. The purpose is to describe how external actors (the MAC and the radio) try to achieve a goal by using the system (the PHY) from a black box point of view.

It seems straightforward to identify quickly the several blocks of the PHY shown on Figure 10.5, such as the two modems and the interface to the lower (RF) and upper (MAC) levels. But doing so at a too early stage encourages the structural separation of the modems and fails to identify a tight functional link imposed by the requirements. A use case analysis, a key activity which will be detailed in the next paragraphs, reveals instead that the PHY must detect the type of frame when receiving a valid signal in the 2.4 GHz band. The modems both try to synchronize on the incoming signal and recognize their respective preamble. This operation may be coordinated by an arbiter taking a decision based on the results from both modems. An early decomposition of the system restricts the design space exploration and can preclude the investigation of design alternatives. An architecture actually realizes the use cases (Figure 10.6) and is proposed after the use case analysis. The system's behavior is understood during the use case analysis without introducing design details. Nevertheless, constraints coming from design aspects (for example, an available platform) may already be present from the beginning of the system's development, as in this example.

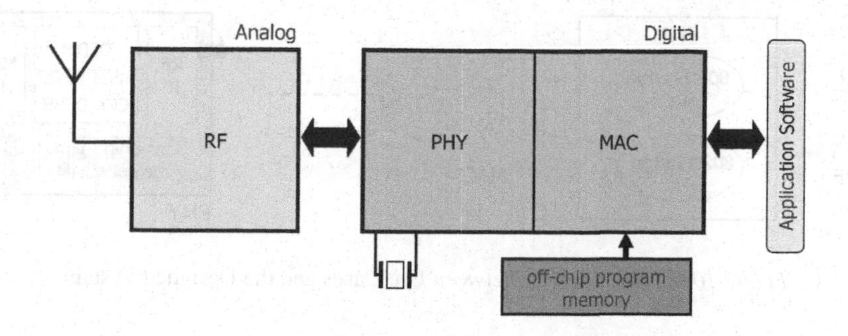

Figure 10.3. High Level View of the Architecture

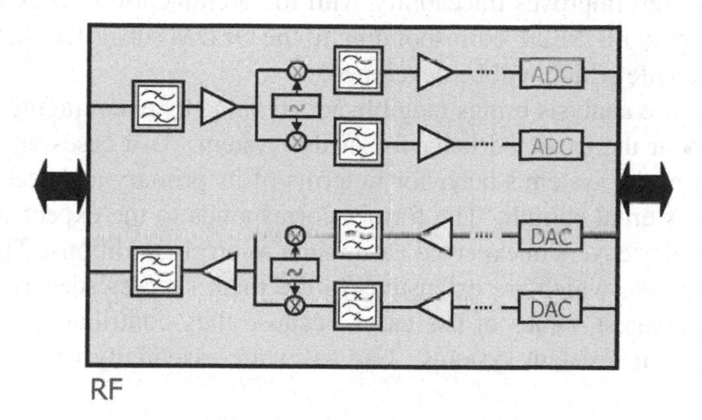

Figure 10.4. Overview of the Radio Frequency Part

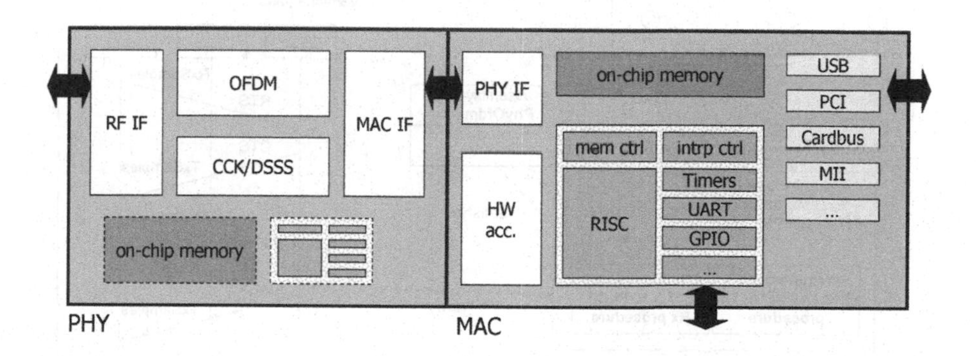

Figure 10.5. Overview of the Physical and Medium Access Control Layers

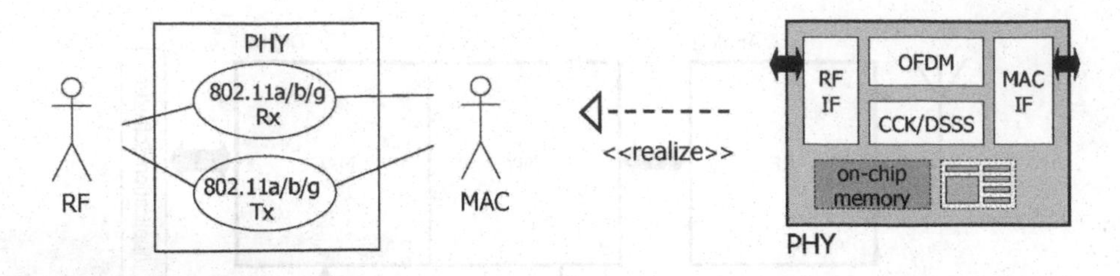

Figure 10.6. Relationship Between Use Cases and the Designed System

Before the use case analysis starts, requirements can be structured by means of the SysML requirements diagram (Figure 10.7). As indicated in Section 10.2.3, this step improves traceability with the architecture (represented here by the assembly `PhyOFDM` corresponding to the OFDM subsystem in the PHY) and the test suite which will be developed.

The use case analysis brings tangible added value by encouraging thorough thinking about the expected behavior of the system. Use cases are a textual description of the system's behavior in terms of its primary and secondary responses to external stimuli. The former corresponds to the expected reaction whilst the latter covers unexpected cases such as error conditions. These alternative behaviors, which are extensions to the main success scenario, provide one of the greatest values of use cases because they contribute to designing robust and fault resistant systems. Use cases are essentially text despite that

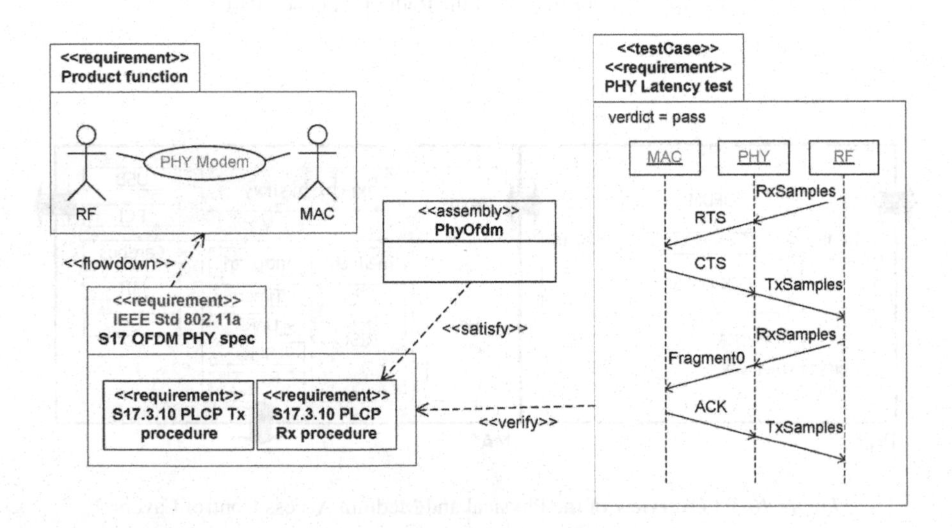

Figure 10.7. Requirement Diagram and Traceability to Design and Test

Use Case – Tx 802.11g OFDM

scope: PHY system

trigger: MAC issues a Tx request (2.4 GHz band).

preconditions: CCA idle

postconditions: PHY available for a new Rx or Tx request

success guarantee: Frame mapped to a stream of samples transferred to the RF for transmission in the 2.4 GHz band

main success scenario:

1 The MAC provides the RATE, LENGTH, and SERVICE parameters to the PHY. The PHY calculates the number of tail and pad bits.

2 The PHY generates the SIGNAL field, encodes it with a convolutional code at R=1/2, and maps it onto an OFDM symbol with BPSK modulation.

3 The PHY configures the RF to use the 2.4 GHz band, and starts sending the training symbols to the RF.

4 The MAC provides the data to the PHY at a maximum average rate of 54 Mbps.

5 The PHY scrambles and encodes the data with the rate specified by RATE.

...

Steps 4–9 repeat until the end of the frame.

10 The PHY notifies the MAC when it transmits the last sample to the RF.

extensions:

1.a The RATE parameter is not valid.

1.a.1 The PHY notifies a Tx error to the MAC and resets its state.

4.a The MAC interrupts its data transfer

4.a.1 The PHY aborts transmission to the RF and resets its state.

...

Figure 10.8. Example of Use Case

use case diagrams may help to organize use cases whilst behavioral diagrams such as sequence diagrams can illustrate them visually.

Although use cases provide a powerful modeling instrument, they require skill and experience. Besides the original definition from Jacobson[6], numerous variations have been proposed [93] and differ in the degree of formalism, the organization between use cases, the relationship with respect to scenarios, and the modeling capabilities of use cases. Cockburn [42] recommends writing goal oriented semi-formal text organized in a semi-formal structure, which is an interesting compromise between the need for clear communication and the drawbacks of over-abundant formalism.

As an illustration, the use case of the transmit procedure for OFDM modulation in the 2.4 GHz band may look like Figure 10.8. This example illustrates

[6]"A use case is a sequence of transactions in a system whose task is to yield a measurable value to an individual actor of the system." [102]

a remarkable benefit of the use case analysis: the possible paths in the use case give rise to a collection of scenarios which help to identify the tests to apply later on the designed system. The impact of the use case analysis therefore spans the complete development until the test of the prototype. Use cases contribute to the principle of testing early and often. The added value of use cases as a means for regular, systematic, and effective testing depends, however, on the rigor and formality of the use cases.

Use cases are discovered by identifying the scope of the system, the external actors, and their goals. In the SoC context the term actor should be interpreted in a wide sense. It corresponds to any object located outside the system considered and interacting directly with it, ranging from other large systems (such as the MAC and the RF in the present case) to a simple time trigger, for example.

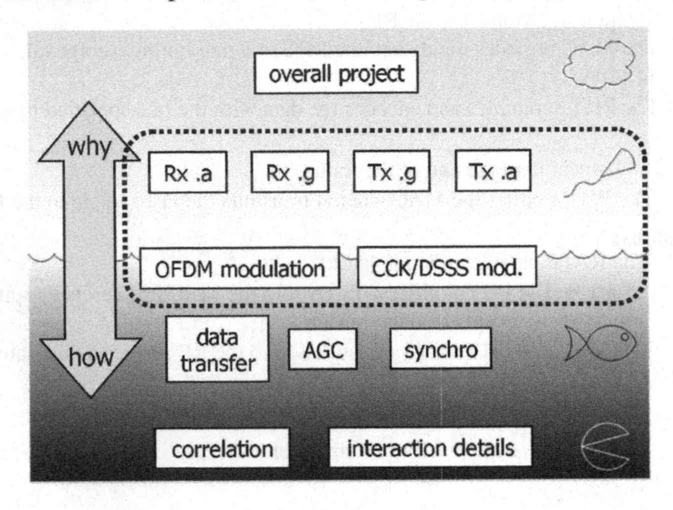

Figure 10.9. Use Case Levels

As shown in Figure 10.9, use cases can be structured according to their goal level [41] instead of using the object oriented relationships between use cases as defined in UML. Higher levels provide summary use cases and tend to answer the question why these use cases are of interest, whilst lower levels correspond to sub-functions closer to the question of how the functionality is done. Valuable use cases are located mainly inside the dotted line at the 'sea-level'. The scope and the level of use cases are key practical aspects to keep in mind during the use case analysis.

Limitations, Pitfalls. It is important to underline several misconceptions and limits of use cases. First, the distinction between requirements and use cases is a common point of misinterpretation. Use cases provide a technique for analyzing specifications, mitigating the functional risk caused by misunderstanding the requirements. Therefore use cases are mainly equivalent to the *functional requirements*, since they detail what the system must do. Use cases provide limited support for capturing the *non-functional requirements* or *quality of service* (QoS) *requirements* in the large sense, such as constraints related to time, power consumption or computational accuracy. This type of requirements is associated with the performance risk, caused by insufficient evaluation of the system performance throughout the design. *Primary* QoS requirements, such as the packet error rate tolerated by the IEEE WLAN standard, can be represented in SysML requirements diagrams and traced to the designed SoC. *Derived* QoS requirements, such as a limited processing time, are the consequences of primary QoS requirements and design constraints. The degree of satisfaction of the system in terms of its QoS requirements can be evaluated by means of executable models of the SoC, using for example C/C++ based languages such as SystemC [219, 64, 74]. This approach avoids postponing and ignoring crucial parts of the system analysis. As the results are only accurate to the level of the model of the system, dedicated attention must be paid to the assumptions and the abstraction level of the model.

Second, system architects often have exaggerated expectations with use cases. In particular, it should be underlined that use cases describe the behavioral requirements but do not lead straightforwardly to an implementation. Applying a functional decomposition of the system structure with the help of the use cases is a mistake. Instead, a proposed implementation must be validated by means of the scenarios from the use cases.

Finally, numerous questions such as the determination of the completeness, the correctness, the consistency of use cases, or the appropriate level of details, commonly arise while writing use cases. The reader is referred to the available literature for practical hints and solutions [41, 112].

10.4.3 From Use Cases to Architecture Modeling

Realization of the Top Level Use Cases. Contrary to a waterfall approach, the evolution from the use case analysis to the actual design occurs smoothly. Elements of a possible architecture are gradually introduced as soon as the major use cases have been developed while use cases are further refined. The system is opened up, broken down into large scale subsystems to which clear responsibilities are assigned (Figure 10.10). Referring to the introductory example of Section 10.4.2, it should be stressed that these main components are identified only now. This draft architecture is then validated against the top level use cases. The use case analysis is therefore a critical step, as it drives the complete design process. Unchecked errors or unexpected behaviors risk propagating through the project with dramatical consequences.

Figure 10.11 presents a sequence diagram corresponding to the use case for the reception in the 2.4 GHz band, in which the system is now seen as a white

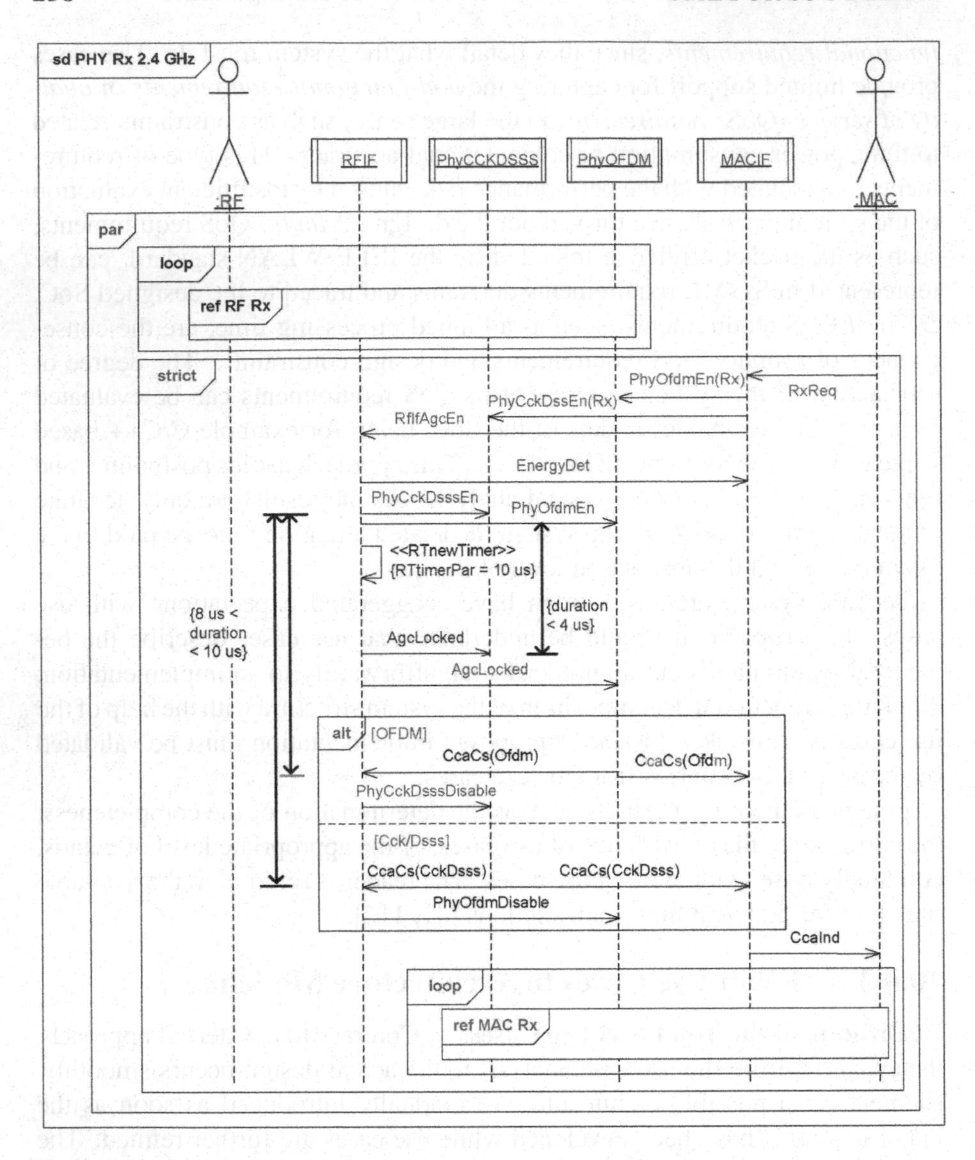

Figure 10.10. PHY Top Level Rx 2.4 GHz Sequence Diagram

box view. As shown in the referenced sequence diagram (Figure 10.11), the PHY receives samples from the RF at a rate dictated by the ADC, in the present case 40 MHz. Owing to the different data rates supported by the OFDM and the CCK/DSSS modulations, the two modems work at different frequencies, 20 and 22 MHz. The block interfacing the RF performs the appropriate rate conversion. Upon request from the MAC, the PHY listens to the channel and tries to detect an OFDM or a CCK/DSSS signal. If a signal is present the

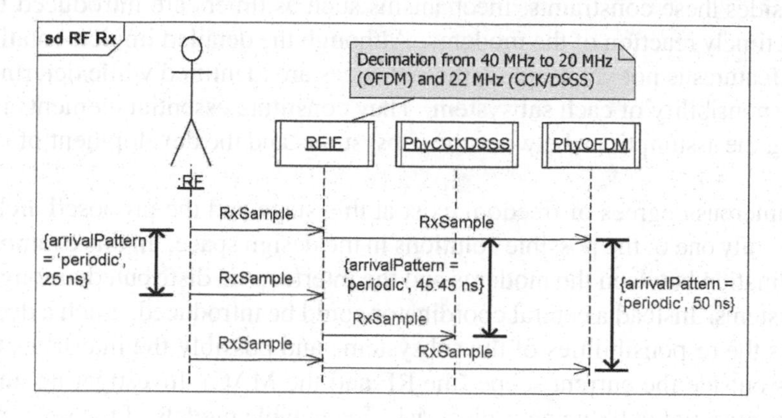

Figure 10.11. PHY Sequence Diagram for the Reception of Samples from the RF

RFIF detects that the channel is busy and enables both modems. The RFIF adjusts the gain so that the amplified signal can be processed by the modems. These attempt to recognize specific characteristics of the OFDM or CCK/DSSS preamble initiating each frame. Once one of the modems detects the pattern it expects it notifies the RFIF, which consequently disables the other modem. Received samples are then demodulated and transferred to the MAC until the end of the frame, as indicated by the loop fragment.

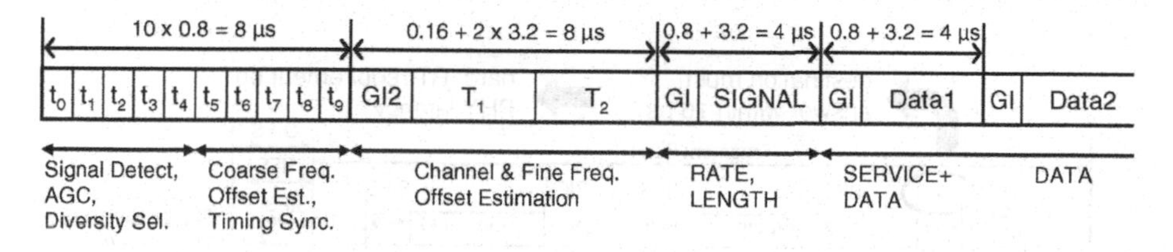

Figure 10.12. OFDM Preamble

Figure 10.12 represents the preamble which constitutes the beginning of an OFDM modulated frame [96] and imposes several hard real-time constraints. The Automatic Gain Control (AGC) functionality should, for example, be performed in 4 μs at most, whilst the modems must detect the preamble type in a window time lying between 8 and 10 μs. Sequence diagrams can be annotated with real-time constraints expressed as boolean expressions placed within curly braces. However, UML does not specify the precise syntax nor the semantics of such timing markers. Even though the Real-Time Profile (RTP) provides a set of standard stereotypes for expressing time related aspects, it does not solve this issue completely. Each particular development process or methodology must therefore clarify the remaining ambiguities.

Besides these constraints, mechanisms such as timers are introduced to ensure a timely reaction of the modems. Although the detailed implementation of these features is not yet decided, these devices are identified while determining the responsibility of each subsystem. They constitute essential elements in satisfying the assumptions between the subsystems and the development of robust SoCs.

Numerous degrees of freedom exist at this stage and the proposed architecture is only one of the possible solutions in the design space. In this example the coordination between the modems and the interfaces is distributed amongst the subsystems. Instead a central coordinator could be introduced. Such a decision affects the responsibilities of the subsystems and possibly the interfaces to the actors outside the current scope (the RF and the MAC). Investigating alternatives is essential to reducing project risk. Executable models of the system here play a decisive role because they help verify early the performance of design alternatives in a limited amount of time. Metrics specific to the particular application must be defined to assess the performances of these design alternatives. In this example the IEEE standard imposes a worst case answer time of 16 μs, the duration of the short inter-frame space (SIFS, see Figure 10.13), and design alternatives will have different time margins with respect to this constraint. However time is not the only criterion. Other aspects such as power consumption must be taken into consideration despite the possible difficulty in accurately estimating these factors. This question is further addressed in Section 10.4.4.

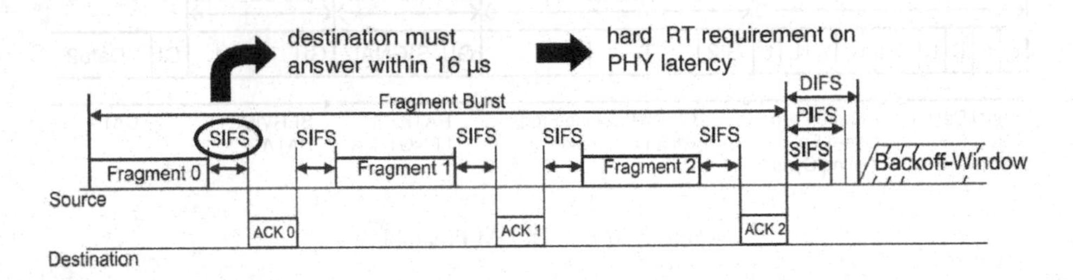

Figure 10.13. Maximum Allowed Inter Frame Times

Refinement of the Top Level Use Cases. The top level use cases are refined while design elements are gradually introduced: the use case analysis is performed recursively at the subsystem scope in a similar way as for the top level. The subsystem use cases are derived from the top level requirements and the architecture decomposition of the system. For instance, the activity diagram in Figure 10.14 is a more detailed view of the MACIF behavior resulting from a use case analysis on this subsystem.

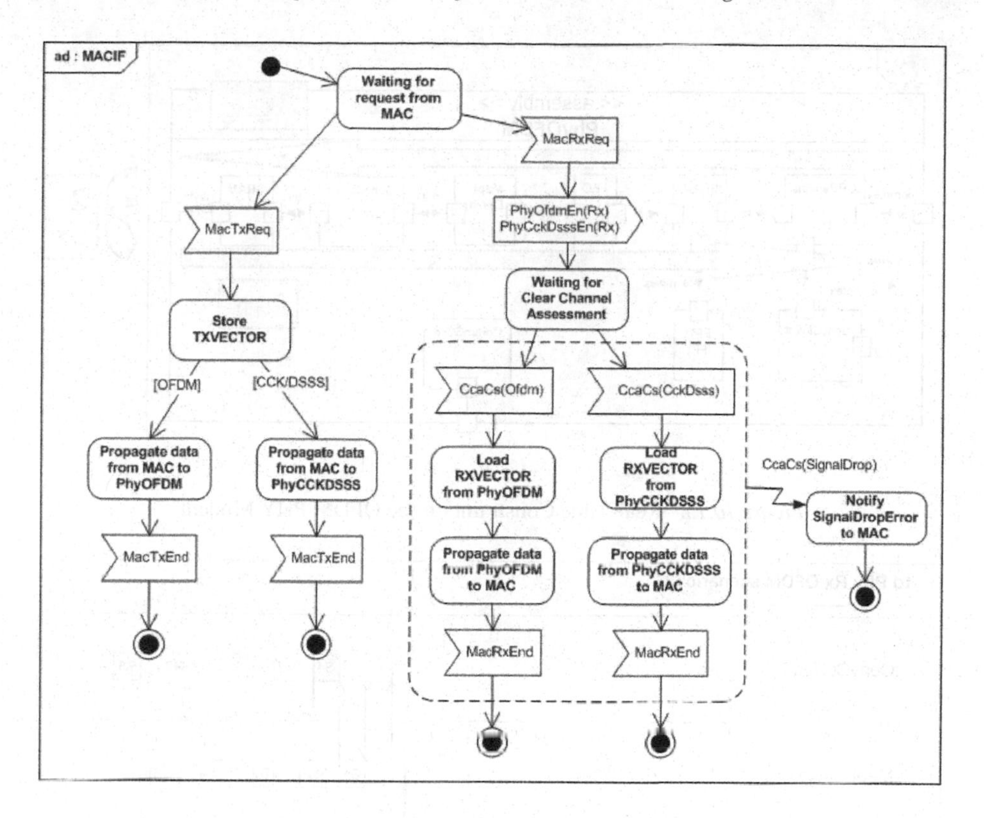

Figure 10.14. MAC Interface Behavior

As a result the transition between the use case analysis and the design effort occurs smoothly and follows the principle of iterative development processes. The algorithmic work proceeds in parallel and follows conventional lines.

The degree to which design details should be introduced varies case by case and is a tradeoff between two opposite concerns: whereas it is a mistake to bring design details too early, the architecture must be validated with enough precision to avoid costly redesigns. Hidden requirements are discovered while moving further towards design. As an example the severe constraint on the reaction time of a terminal (Figure 10.13) has a direct impact on the maximum latency of the PHY. Assuming a MAC reaction time of 2 μs and a preparation time of 2 μs in the PHY to start the generation of the preamble in the Tx direction, the PHY must have a latency smaller than 12 μs in the Rx direction. This requirement propagates further down at the level of the different subsystems within the modems. The time budget represented informally in Figure 10.15 for the OFDM modem is more accurately represented by a time diagram (Figure 10.16).

Figure 10.17 shows how the timing budget is specified at the detailed level of the FSM describing the synchronization of the OFDM modem.

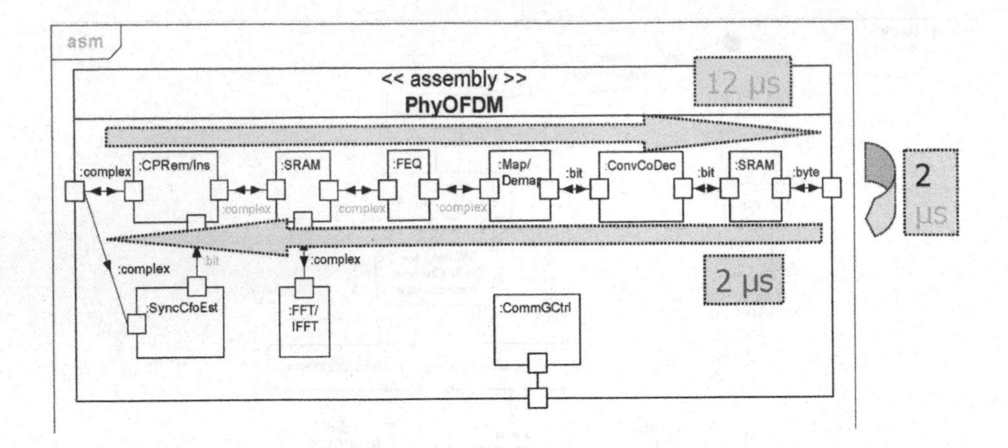

Figure 10.15. Real-Time Constraint on the OFDM PHY Modem

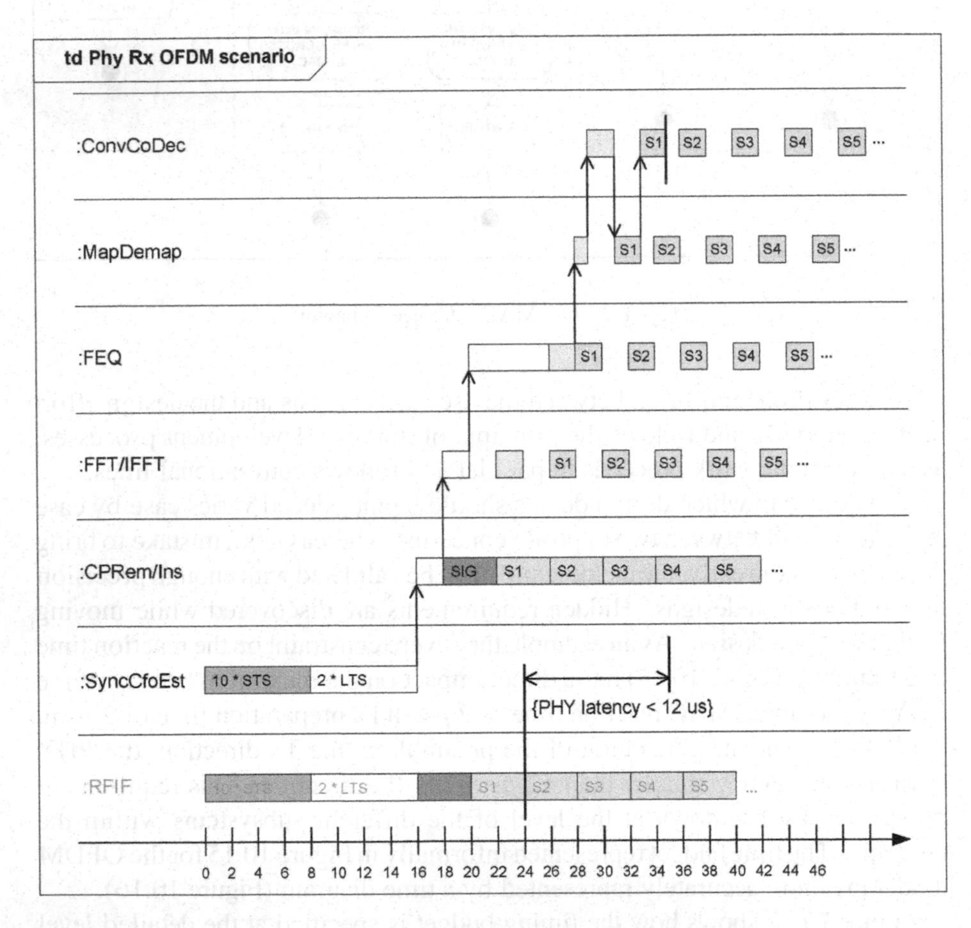

Figure 10.16. Time Diagram for the OFDM PHY Modem

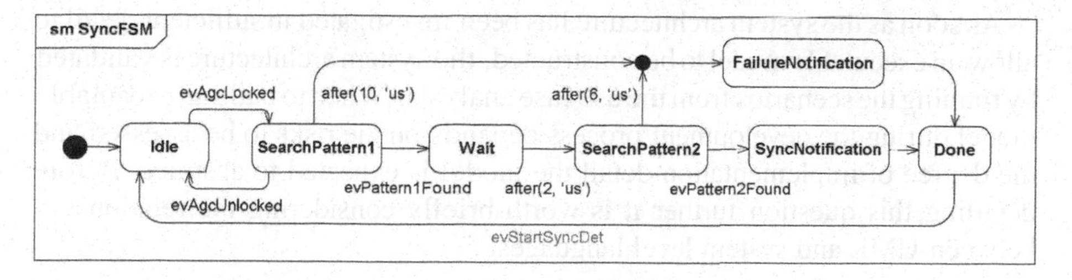

Figure 10.17. Time Budget of the OFDM Synchronization FSM

A SysML assembly describes the system in terms of its structure and the data flow, but does not imply any particular implementation domain and partitioning between hardware and software. By analyzing the system architecture further and identifying the main blocks and associated time budgets it becomes clear that several parts, such as the synchronization, have very high processing demands which a software implementation cannot provide. Neither can other parts be implemented on a general purpose processor, but a tradeoff between the performances of dedicated hardware and the flexibility of software can be obtained by means of custom processors like application specific instruction processors (ASIPs). Memories are introduced to relax the timing constraints between the processing blocks. Further questions related to the partitioning between hardware and software or the number of processors are outside the scope of this chapter. At the end of this step one or several partitioning solutions arise and can be investigated further. Figure 10.18 depicts one of the possible choices for the present case study.

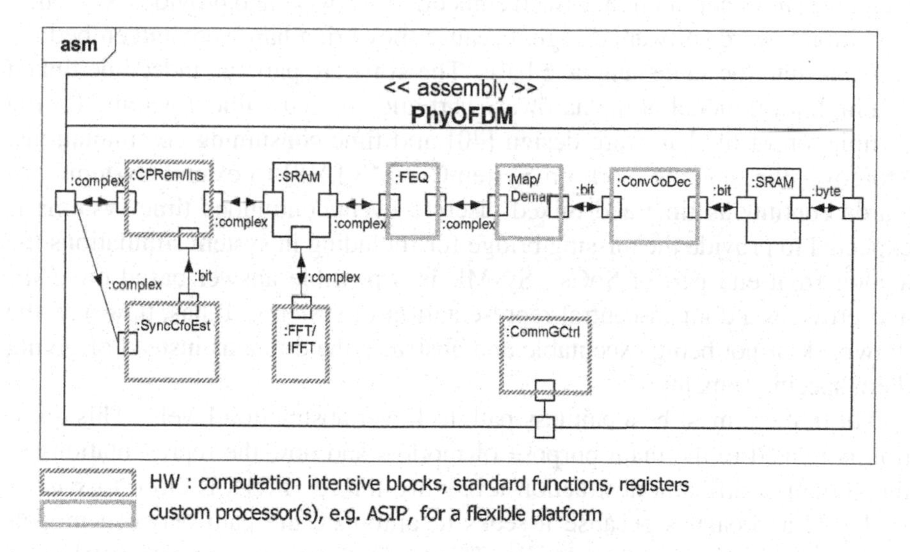

Figure 10.18. Hardware/Software Partitioning of the OFDM Modem of the PHY

As soon as the system architecture has been investigated in sufficient detail to allow an executable model to be constructed, the system architecture is validated by running the scenarios from the use case analysis. When to build an executable model during the development process depends on the risks to be assessed and the degree of implementation detail the model is expected to abstract. Before detailing this question further it is worth briefly considering the relationship between UML and system level languages.

UML and SoC Languages at System Level. Several languages have been developed in recent years to support higher abstraction levels than RTL. Proposals for system level languages are based on extensions of existing software or hardware languages, such as C (SpecC, Handel-C), C++ (SystemC, Cynlib), VHDL or Verilog (SystemVerilog), or new languages created specifically for system level design (Rosetta) [82]. The different flavors of executable UML fall in the latter category as they rely on a tool specific action language which is platform-independent, although some code belonging to one of the languages above is eventually generated by a model compiler provided that mapping rules are defined for that specific language. In any case there are a number of essential requirements a language must fulfil at system level.

First, a language can only be executed if its syntax and semantics are unambiguous. As a consequence UML itself is not enough and must either have its semantics clarified or be associated with an unambiguous executable language.

Second, a language at system level must support abstraction and be able to describe heterogeneous implementation domains (hardware/software, analog/digital) and models of computation (MOC). In this respect C/C++ based languages have common roots with embedded software and provide a smoother transition toward software design, because they bring hardware and embedded software into the same language base. The software part can indeed be linked to a high level model of the hardware part and verified without waiting for the completion of the hardware design [40] and time consuming co-simulations. Moreover the ongoing work on SystemC-AMS [218] to extend SystemC towards continuous time and mixed discrete event/continuous time systems is expected to provide the missing bridge for including in system simulations the analog front-end part of SoCs. SysML is a possible answer based on UML and provides a domain-neutral representation of systems. It has, however, the drawback of not being executable and abstracts the domains instead of giving them specific semantics.

Third, there must be a painless path to lower abstraction levels. This question is related to the main purpose of models and how the representations of the system at different abstraction levels are linked. MDA is close to *synthesis*-based approaches because it seeks to automate the path between models and abstract implementation details. This typically imposes several restrictions on the models and the language. According to this view models are a representation of the system at a higher level but still have a defined link with the implementation. On the other hand, *verification*-oriented solutions consider models as a risk-mitigating tool and emphasize the role of test bench reuse, testing the system at a lower abstraction level (such as cycle-accurate RTL) using higher level test benches (e.g. based on SystemC and SCV, the SystemC Verification library). Models are used here mainly as an abstract representation of a system in order to gain a thorough understanding of the system before starting the design and testing it early. Furthermore, models act as a reference for further design steps and reduce the problem of creating synthesizable models. However, this creates a gap between models and implementation that must be carefully compensated by thorough testing. For reasons explained in the introduction, severe implementation constraints may restrict the application of MDA to SoC design. In addition, model-driven approaches focusing on the verification of the functionality and the performance of a system can take benefit from languages which support several abstraction levels at the same time, such as SystemC. This feature allows gradually refining parts of the model in order to reduce the gap between transaction level models of the system and its RTL description [219], and supports the principles of iterative development processes in which the system designed is gradually refined. Object-oriented languages naturally complement SoC methodologies which involve UML, since the link between models expressed in OO languages and their UML representation follows principles similar to the application of UML to classical software artifacts. Figure 10.19 shows how the SystemC model of the OFDM PHY subsystem can be documented using UML. However, UML is relevant with non-OO languages as well, as is indicated by early works of VHDL generation from UML models for verification [130] or more recent efforts for synthesis purposes [19, 184].

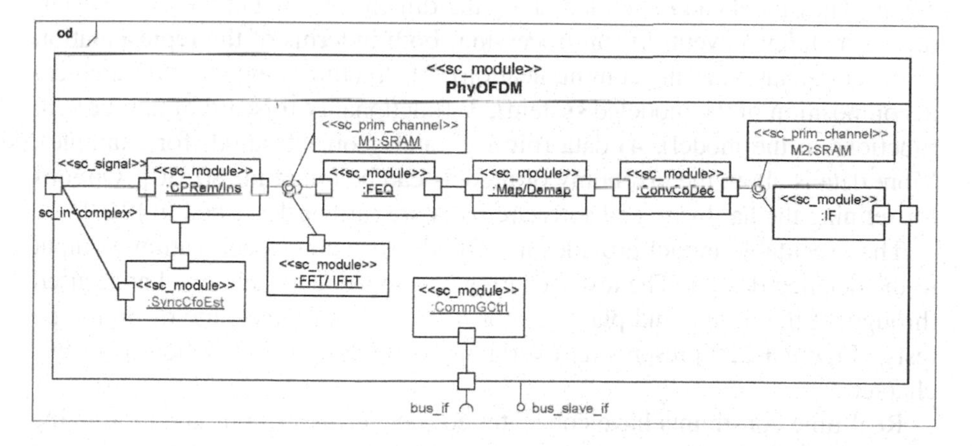

Figure 10.19. UML Representation of the SystemC Model of the OFDM PHY Modem

As these examples illustrate, requirements analyzed with the help of UML may be implemented without using OO.

The bottom line is that UML can be complemented with several SoC design languages. This is required when using standard UML because it provides only a notation and is not executable. We will therefore close this section by considering SystemC as the target language to model and assess the system architecture.

Executable Models of the Architecture. The purposes of an executable model of the architecture using, e.g., SystemC can be summarized as follows. First, it is constructed to validate completely the system's behavior, because it avoids the ambiguity of paper specifications. Second, the model is meant to capture essential characteristics of the system such that its performance can be quantified. Although some information is lost when raising the abstraction level, the impact of limited precision and fixed point effects on intensive digital signal processing hardware parts can for example be quickly assessed. Third, the model abstracts the description of the hardware and provides a platform for an early development of the embedded software part of the design. Last, but not least, an executable model acts as a reference and tool for test generation as input to the detailed hardware design. The SystemC executable model adds precision to the scenarios generated from the use case analysis and provides strong cross verification capabilities between the model of the architecture and the RTL description. Although SystemC can reach RTL level and be used for synthesis purposes, doing so does not take advantage of the major assets of SystemC and does not correspond to its mainstream use [51].

The ability to model the system at different levels of abstraction is vital to support an iterative development process, since it allows one to gradually construct more refined versions of the system model with an increasing amount of detail. The model can be refined along the dimensions of 1) time (cycle accurate or high level event-driven processing, both in terms of the representation of the computation as the communication), 2) structure (degree of hierarchical decomposition of the modeled system), 3) functionality (discard of non-critical functions in the model), 4) data (bit accurate or object based, for example). More details about the taxonomy of models can be found in [80, 30]. Options for refining the hardware and software parts are further described in [219].

The executable model provides a verified executable specification as input to the detailed design. The test scenarios developed at system level are reused throughout the design and play a major role in defining the suite of prototype tests. Figure 10.20 presents an overview of the design flow followed in this chapter.

Real-time constraints have concentrated a major part of our attention in the previous sections. As power consumption is becoming a technological issue

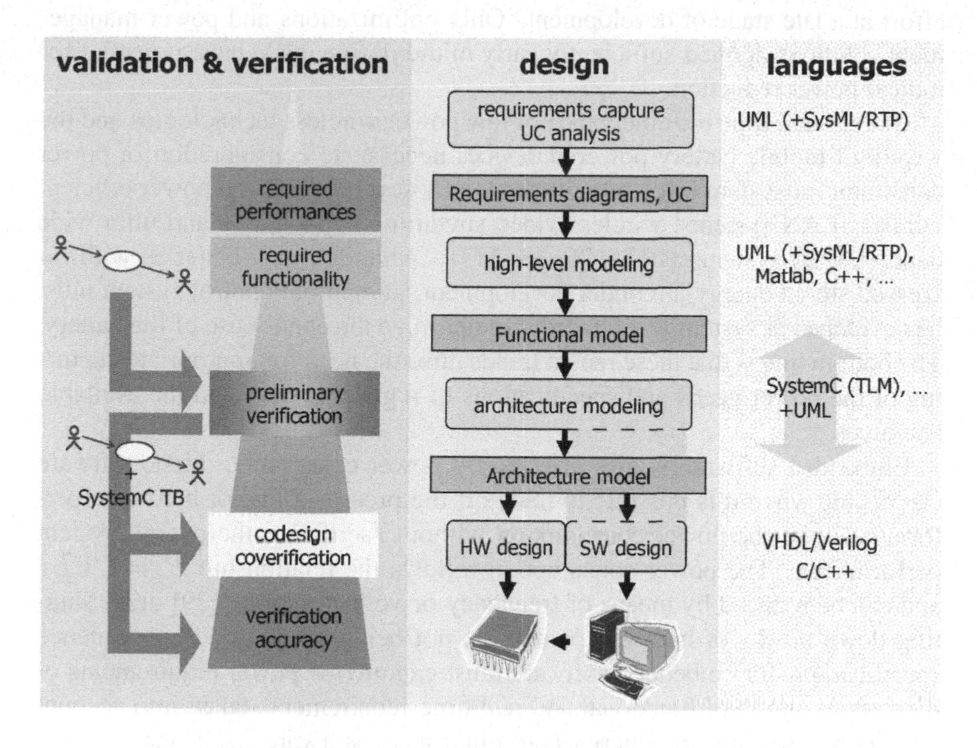

Figure 10.20. UML/SystemC Based Design Flow

of crucial importance, UML should also facilitate the development of power efficient systems. In the remainder of this chapter we look further into this question with the help of the same case study.

10.4.4 Power Efficient System Design: Can UML Help?

Need for Power Aware System Design. Current works investigating the application of UML to SoC design seek to bridge the gap between heterogeneous domains but do not take into account power consumption, a rising technological concern of major consequence [180, 39].

As the feature size of the VLSI circuit technology decreases towards the deep submicron range, device physics increasingly threatens classical development approaches. These separate digital design aspects as much as possible from manufacturing issues and concentrate design efforts on area minimization and timing analysis. However, severe power constraints emerge and are now imposed through the entire design flow, motivating power management solutions at each level of abstraction. Numerous power optimization techniques ranging from architectural to circuit level are available [33]. However, the power saving capabilities of these methods are limited and concentrate the power optimization effort at a late stage of development. Only optimizations and power management solutions applied sufficiently early in the design cycle have potential for radical power reduction.

At the same time the emergence of low power wireless technologies and the wealth of mobile battery powered devices necessitate consideration of power constraints at system level, as is illustrated by several works on power efficient wireless LAN systems, wireless video streaming applications and ultra wide band (UWB) systems [107, 176, 38, 117]. Although new batteries with increased stored energy are under development, an optimal control system must be developed at system level in order to optimize the energy use of the battery. The bottom line is that these recent trends raise the need for appropriate capturing of the power needs and specifications of a given system and the available resources.

Embedded software is also affected by power constraints. First, software can decide when it is possible to trade off the processor and/or hardware performance and the power consumption without sacrificing the overall system performance. The power consumption follows the relationship $P = CV_{dd}^2 f$ and can be reduced by means of frequency or voltage scaling [29] or by shutting down blocks of hardware which are not being used. As a consequence specifications for embedded software must capture the power requirements of the system on top of functional and real-time requirements taken into account until now. Second, the energy consumption of software itself and the processor on which it is running becomes a relevant concern and has attracted attention [216, 198, 23, 60].

In a word, recent technological evolutions both in terms of new applications and underlying implementation technology require power aware design flows. The long term success of UML in the domain of SoC design does not only depend on its capability of supporting the design of complex systems but also on its potential to take into account the specificity of power constrained systems.

Role of UML in Power Efficient System Design. As underlined in the previous paragraphs, power constraints must be taken into account early in the design flow. The approach presented in this chapter allows the specification of power needs at system level in order to identify power management solutions early in the design. The benefits of SysML in terms of improved traceability of the requirements towards the actual design are applicable to power constraints.

Furthermore, the stereotypes defined in the RTP profile can be applied beyond the context of time constrained systems, as is illustrated by following examples. A battery can be modeled as a passive resource with specific attributes, such as its energy density or discharge characteristics (Figure 10.21). Dynamic power management solutions which aim at extending the battery lifetime are equivalent to the concept of a resource manager applying a resource control policy to the

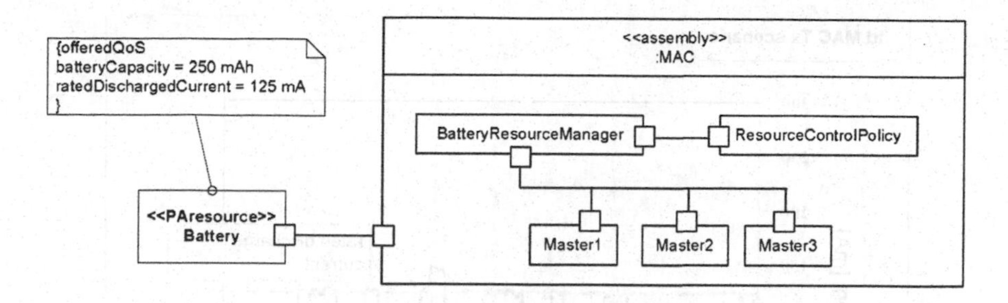

Figure 10.21. Applying the RTP to Model Resources and Battery Access Policy

battery. The specificity of the battery control policy is application-dependent and out of the scope of the RTP. It could be based, for example, on an algorithm offloading tasks between terminals [176], or combined with the time scheduling algorithm such that power hungry tasks do not overlap in time [107].

But the RTP provides only a limited answer. An energy conscious policy such as in [176] considers the energy dissipation cost of the communication when dispatching the heavy workload from a mobile host to an AC powered server. The message exchange in such a case needs proper annotation to capture the energy cost associated with the communication and the offloaded computation, where the RTP falls short. A similar conclusion can be drawn in the context of power aware routing algorithms at MAC level for Ultra Wide Band (UWB) communications [117]. Such examples raise again the issue of standardization and semantics of the annotations.

Even if designers have the means of capturing the power constraints of a SoC in their UML models, the most important question is the ability to support power consumption analysis at system level. Power optimization techniques can be enhanced with the visual means of UML 2. Figure 10.22 is closely related to Figure 10.21 and further details the example [107] mentioned above. In this case the tasks at the MAC level of a wireless LAN system are scheduled such that the current peaks caused by an overlap of power demanding tasks are avoided. In this way the discharge profile becomes more battery friendly. Although power is traded versus time in a more subtle way than the classical dynamic frequency scaling approach, the data rate may decrease. Even though the benefit is not apparent at first sight, the end result is an improved battery capacity and extended lifetime. A time diagram provides a visual means of easily representing the overall activity and the effect of the block scheduling on the total power consumption.

However, power analysis techniques require executable models in order to provide accurate results. A *static* analysis of annotated UML models, as suggested by the RTP, is not adequate for power estimation. An executable HW/SW

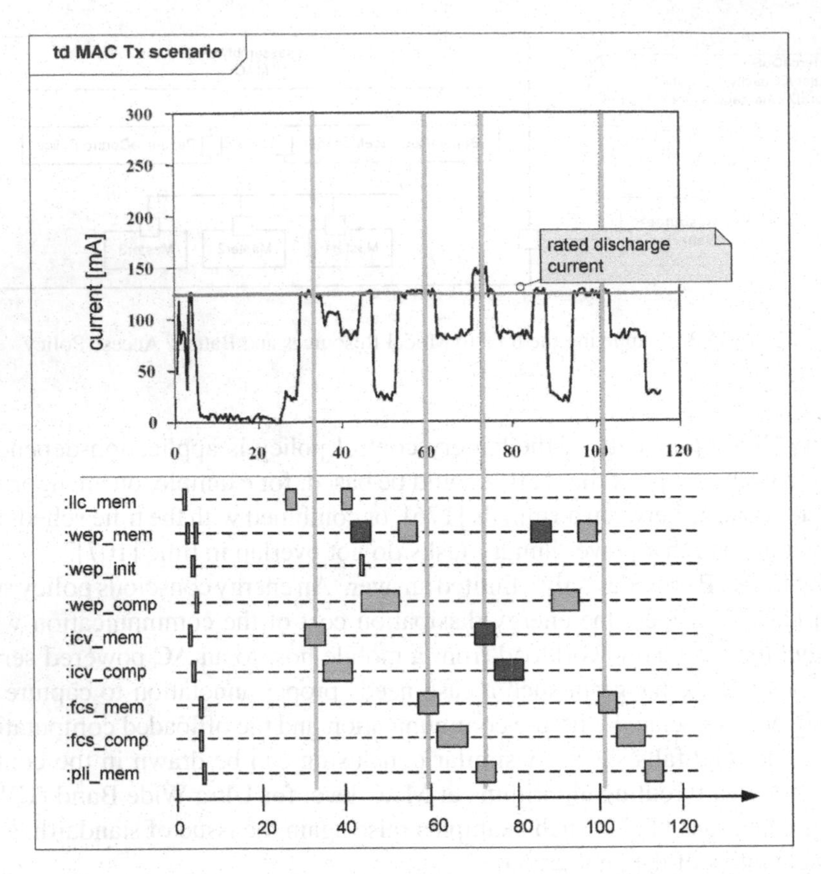

Figure 10.22. Battery Friendly Battery Discharge Profile: Unmodified (Light Grey) and Delayed (Dark Grey) Tasks

cosimulation environment is required such that a *dynamic* or a *mixed* performance analysis can be carried out. The example above corresponds to the latter situation [108]. A detailed simulation is performed once in order to extract the power consumption figures of selected blocks. This characterization step is followed by the static generation of several power traces depending on how these blocks are scheduled. A dynamic performance analysis relies solely on simulation traces. It is therefore slower but can support the analysis of more complicated scenarios. Instead of generating the power consumption curve for a single scenario as above, the power consumption figures extracted in the characterization step can be included in the SystemC executable model to simulate more scenarios. A time diagram can then be generated based on the simulation results. The bottom line is that UML can provide a visual aid and complement existing power analysis techniques, but again requires an executable environment.

Still the question of the power estimation, how to obtain accurate power figures, is the critical aspect of any power analysis technique. The importance of this issue must be stressed although it is beyond the scope of this chapter. One possibility is to characterize and extract parameters to be used at higher levels of abstraction, as shown above. Another solution is to synthesize a model at a lower abstraction level, perform the power estimation at that level by means of activity files from simulations, and bring the results back to the user. This approach is classically implemented in conjunction with RTL synthesis tools as well as more recent behavioral synthesis tools starting from VHDL or C/C++. Obviously UML is particularly an interesting candidate to provide a graphical aid in the latter case. The main issue remains the accuracy of these power estimations methods at system level, which still need to be assessed against several concrete cases and implementations.

A final remark should be made regarding power aware embedded software. The RTP Profile focuses on the capability of capturing time information essential for a particular analysis technique such that predictions can be made about the real-time performance of a system under different conditions. However, several works related to energy aware scheduling algorithms could suggest further enhancements of the profile [233, 230]. Such algorithms optimize timeliness and energy consumption in a unified way. The RTP does not allow scheduling techniques which take into account more constraints than only time requirements, and considers only time as the major non-functional aspect. Stronger support for power aware embedded systems is required.

10.5 Conclusions

The recent advances in UML and SysML present a valuable milestone for the application of UML to modern chip design. Further formalization and standardization are needed. In addition, increasing attention must be paid to power aware design flows in order to consolidate the applicability of UML to SoC design in the future. It is essential to remember that the standard UML is only a notation. UML does not solve the difficulty associated with systems analysis but provides means for efficient communication between the project stakeholders. Hence a sound development process which suits the peculiarities of SoC design is necessary to complement the use of UML for SoC design. Furthermore, executable models are required for early and detailed analysis of the system performances. Several options are possible: besides xtUML the standard UML can be associated with existing system level languages, in particular C/C++ based languages. There is ample room for discussion on the pros and cons of different SoC development processes using UML and how these consider the role of models. Irrelevant to the specific characteristics of the particular adopted methodology, it is beneficial to extend SoC engineering from its focus on hardware and software technologies to an engineering discipline of system development. A multi-disciplinary approach is crucial to coping not only with macroscopic aspects, such as increasing product complexity and reduced 'time to market' requirements, but also the consequences of the physical aspects of SoC, in particular issues related to power consumption.

References

[1] Action Semantics Consortium - UML Subcommittee Webpage. `www.kabira.com/as/home.html`

[2] J. Alderman, M. London, S. Sorensen, and D. Yang. *SHOOT-EMU—Software/Hardware Object Oriented Tool for Executable Modelling*. Fourth Year Group Project Report, Department of Computation, UMIST, June 2003.

[3] P. Alexander, R. Kamath, and D. Barton. System Specification in Rosetta. In *IEEE Engineering of Computer Based Systems Symposium*, Edinburgh, Scotland, 2000.

[4] M. Ardis. A Framework for Evaluating Specification Methods for Reactive Systems: Experience Report. In *IEEE Transactions on Software Engineering*, 22(6), IEEE Press, 1996.

[5] M. Awad, J. Kuusela, and J. Ziegler *Object-Oriented Technology for Real-Time Systems: A Practical Approach Using OMT and Fusion.* Prentice-Hall, Eaglewood Cliffs, NJ, USA, 1996.

[6] ARM Limited. AMBA AHB Cycle Level Interface Specification, 2003.

[7] D.I. August, K. Keutzer, S. Malik, and A.R. Newton. A Disciplined Approach to the Development of Platform Architectures. In *Microelectronics Journal*, 33(11), November 2002.

[8] J. Axelsson. Real-World Modeling in UML. In *13th International Conference on Software and Systems Engineering and their Applications*, Paris, France, December 2000.

[9] A.T. Bahill and B. Gissing. Re-evaluating Systems Engineering Concepts Using Systems Thinking. In *IEEE Transactions on Systems, Man, and Cybernetics—Part C: Applications and Reviews*, 1998.

[10] F. Balarin, M. Chiodo, P. Giusto, H. Hsieh, A. Jurecska, L. Lavagno, C. Passerone, A. Sangiovanni-Vincentelli, E. Sentovich, K. Suzuki,

and B. Tabbara. *Hardware-Software Co-Design of Embedded Systems: The POLIS Approach*. Kluwer Academic Publishers, Boston/Dordrecht/London, 1997.

[11] F. Balarin, Y. Watanabe, H. Hsieh, L. Lavagno, C. Passerone, and A. Sangiovanni-Vincentelli. Metropolis: An Integrated Electronic System Design Environment. In *IEEE Computer*, 36(4), April 2003.

[12] L. Baresi, F. Bruschi, E. Di Nitto, and D. Sciuto. SystemC Code Generation from UML Models. In Ch. Grimm (ed.): *System Specification & Design Languages*. Kluwer Academic Publishers, Boston/Dordrecht/London, 2003.

[13] A. S. Basu, M. Lajolo, and M. Prevostini. UML in an Electronic System Level Design Methodology. In *UML for SoC Design Workshop at DAC'04 (UML-SoC'04)*, San Diego, CA, USA, June 2004.

[14] Th. Beierlein, D. Fröhlich, and B. Steinbach. UML Based Codesign of Reconfigurable Architectures. In *Forum on Specification and Design Languages (FDL'03)*, Frankfurt, Germany, September 2003.

[15] K. Berkenkötter, S. Bisanz, U. Hannemann, and J. Peleska. Hybrid UML Profile for UML 2.0. In *Workshop on Specification and Validation of UML Models for Real Time and Embedded Systems (SVERTS 2003)*, San Francisco, CA, USA, October 2003.

[16] S.S. Bhattacharyya, P.K. Murthy, and E.A. Lee. *Software Synthesis from Dataflow Graphs*. Kluwer Academic Publishers, Boston/Dordrecht/London, 1996.

[17] L. Bichler, A. Radermacher, and A. Schürr. Integrating Data Flow Equations with UML/Realtime. In *Real-Time Systems*, 26(1), Kluwer Academic Publishers, Boston/Dordrecht/London, January 2004.

[18] M. Björkander and C. Kobryn. Support for Embedded Systems in UML 2.0. In E. Villar and J. Mermet (eds.): *System Specification and Design Languages: Best of FDL 2002*, Kluwer Academic Publishers, Boston/Dordrecht/London, 2003.

[19] D. Björklund and J. Lilius. From UML Behavioral Descriptions to Efficient Synthesizable VHDL. In *IEEE Norchip Conference*, Copenhagen, Denmark, November 2002.

[20] D. Bjorklund, J. Lilius, and I. Porres. A Unified Approach to Code Generation from Behavioral Diagrams. In *Forum on Specification and Design Languages (FDL'03)*, Frankfurt, Germany, September 2003.

[21] C. Bock. UML2 Activity and Action Models Part 3 Control Nodes. *Journal of Object Technology*, 2(6), ETH Swiss Federal Institute of Technology, 2003.

[22] B.W. Boehm. A Spiral Model of Software Development and Enhancement. In *IEEE Computer*, 21(5), May 1988.

[23] A. Bona, M. Sami, D. Sciuto, C. Silvano, V. Zaccaria, and R. Zafalon. Energy Estimation and Optimization of Embedded VLIW Processors based on Instruction Clustering. In *39th Design Automation Conference (DAC'02)*, New Orleans, LA, USA, 2002.

[24] G. Booch, J. Rumbaugh, and I. Jacobson. *Unified Modeling Language User Guide*. Addison-Wesley, Boston, 1999.

[25] G. Boyd. Executable UML: Diagrams for the Future. In *devX.com*, February 5, 2003. www.devx.com/enterpresi/Article/10717

[26] F. Bruschi. A SystemC Based Design Flow Starting from UML Models. In *9th European SystemC Users Group Meeting,* Paris, France, 2004.

[27] F. Bruschi, E. di Nitto, and D. Sciuto. SystemC Code Generation from UML Models. In *System Specification and Design Languages: Best of FDL'02*, Kluwer Academic Publishers, Boston/Dordrecht/London, January 2003.

[28] F. Bruschi and D. Sciuto. *A SystemC Based Design Flow Starting from UML*. In *6th European SystemC Users Meeting*, Stresa, Italy, October 2002. www-ti.informatik.uni-tuebingen.de/~systemc

[29] T. Burd, T. Pering, A. Stratakos, and R. Brodersen. A Dynamic Voltage Scaled Microprocessor System. In *IEEE Journal of Solid-State Circuits*, 35(11), IEEE CS Press, November 2000.

[30] L. Cai and D.D. Gajski. Transaction Level Modeling: An Overview. In *1st IEEE/ACM/IFIP International Conference on Hardware/Software Codesign & System Synthesis,* Newport Beach, CA, USA, October 2003.

[31] M. Caldari, M. Conti, M. Coppola, S. Curaba, L. Pieralisi, and C. Turchetti. Transaction-Level Models for AMBA Bus Architecture Using SystemC 2.0. In *DATE'03 Designers' Forum*, Munich, Germany, 2003.

[32] Celoxica. *Agility Compiler for SystemC Synthesis.* 2005. www.celoxica.com/agility/FlashLoader.htm

[33] A. Chandrakasan and R. Brodersen. *Low-Power CMOS Design*. Wiley-IEEE Press, 1998.

[34] H. Chang, L. Cooke, M. Hunt, G. Martin, A. McNelly, and L. Todd. *Surviving the SOC Revolution*. Kluwer Academic Publishers, Boston/Dordrecht/London, 1999.

[35] W. H. Chen, C. H. Smith, and S. C. Fralick. A Fast Computational Algorithm for the Discrete Cosine Transform. In *IEEE Trans. on Commununications,* vol. COM-25, 1977.

[36] R. Chen, M. Sgroi, G. Martin, L. Lavagno, A. Sangiovanni-Vincentelli, and J. Rabaey. Embedded System Design Using UML and Platforms. In *Forum on Specification and Design Languages (FDL'02),* Marseille, France, September 2002.

[37] R. Chen, M. Sgroi, L. Lavagno, G. Martin, A. Sangiovanni-Vincentelli, and J. Rabaey. UML and Platform-Based Design. In L. Lavagno, G. Martin, and B. Selic (eds.): *UML for Real: Design of Embedded Real-Time Systems*, Kluwer Academic Publishers, Boston/Dordrecht/London, 2003.

[38] K. Choi, K. Kim, and M. Pedram. Energy-Aware MPEG-4 FGS Streaming. In *40th Design Automation Conference (DAC'03)*, Anaheim, CA, USA, 2003.

[39] P. Clarke. Bridging the Divide Between Design and Manufacturing. *Silicon Strategies*, November 2003.

[40] A. Clouard, K. Jain, F. Ghenassia, L. Maillet-Contoz, and J.-P. Strassen. Using Transactional Level Models in a SoC Design Flow. In W. Mueller, W. Rosenstiel, J. Ruf (eds.): *SystemC: Methodologies and Applications*, Kluwer Academic Publishers, Boston/Dordrecht/London, 2003.

[41] A. Cockburn. *Writing Effective Use Cases*. Addison-Wesley, Boston, 2000.

[42] A. Cockburn. Use Cases, Ten Years Later. *STQE Magazine*, 4(2), March-April 2002.

[43] J. Cohn. Technology Challenges for SoC Design: An IBM Perspective. In G. Martin, H. Chang (eds.): *Winning the SoC Revolution*, Kluwer Academic Publishers, Boston/Dordrecht/London, 2003.

[44] CriticalBlue Webpage. `www.criticalblue.com`

[45] A. Cuccuru, P. Marquet, and J.-L. Dekeyser. UML2 as an ADL Hierarchical Hardware Modeling. Technical Report 5166, Institut National de Recherche en Informatique et en Automatique, April 2004.

[46] W. Damm and D. Harel. LSCs: Breathing Life into Message Sequence Charts. In *Formal Methods in System Design,* 19(1), Kluwer Academic Publishers, Boston/Dordrecht/London, 2001.

[47] G. de Jong. A UML-Based Design Methodology for Real-Time and Embedded Systems. In *Design Automation and Test in Europe (DATE'02)*, Paris, France, March 2002.

[48] DOM Webpage. `www.w3.org/DOM`

[49] B.P. Douglass. *Real-Time UML: Developing Efficient Objects for Embedded Systems.* Addison-Wesley, Boston, 2004.

[50] B.P. Douglass. *Rhapsody 5.0 Breakthroughs in Software and Systems Engineering.* White Paper. I-Logix, 2005. `www.ilogix.com/whitepapers/whitepapers.cfm`

[51] Doulos. SystemC in Europe – Current Usage and Future Requirements. Technical Report, Doulos, May 2003.

[52] I. Drost. *Estimation of Execution Probabilities and Frequencies of OO Models*. Diploma Thesis, University of Applied Sciences Mittweida, Germany, 2003.

[53] dSPACE GmbH. *AutomationDesk Guide Release 4.2*. Paderborn, Germany, March 2005. `www.dspace.com`

[54] M.D. Edwards and P.N. Green. UML for Hardware and Software Object Modeling. In L. Lavagno, G. Martin, and B. Selic (eds.): *UML for Real: Design of Embedded Real-Time Systems*, Kluwer Academic Publishers, Boston/Dordrecht/London, 2003.

[55] M. D. Edwards and P. N. Green. Run-time Support for Dynamically Reconfigurable Computing. In *Journal of Systems Architecture: the EUROMICRO Journal*, 49(4-6), Elsevier North-Holland, New York, NY, 2003.

[56] S. Edwards, L. Lavagno, E..A. Lee, and A. Sangiovanni-Vincentelli. Design of Embedded Systems: Formal Models, Validation, and Synthesis. In *Proceedings of IEEE*, 85(3), March 1997.

[57] A. Eliens. *Principles of Object-Oriented Software Development*. Addison-Wesley, Boston, 2000.

[58] J.P. Elliott. *Understanding Behavioral Synthesis.* Kluwer Academic Publishers, Boston/Dordrecht/London, 2000.

[59] H.E. Eriksson, M. Penker, B. Lyons, and D. Fado. *UML 2 Toolkit.* John Wiley & Sons, 2004.

[60] Y. Fei, S. Ravi, A. Raghunathan, and N. Jha. Energy Estimation for Extensible processors. In *Design, Automation and Test in Europe (DATE'03)*, Munich, Germany, 2003.

[61] S. Flake and W. Mueller. An OCL Extension for Real-Time Constraints. In T. Clark and J. Warmer (eds.): *Advances in Object Modelling with the OCL*, Springer Verlag, Berlin, 2001.

[62] C. Flanagan. Extended Static Checking for Java. In *ACM SIGPLAN Programming Language Design and Implementation (PLDI).* Berlin, Germany, 2002.

[63] Forte Design Systems. *Cynthesizer.* 2005. www.forteds.com/products/cynthesizer.asp

[64] Fujitsu. New SoC Design Methodology Based on UML and C Programming Languages. *Fujitsu Electronic Devices News (FIND)*, 20(4), 2002.

[65] Fujitsu Limited, IBM Corporation, NEC Corporation. *A UML Extension Profile for SoC.* Draft RFC Submission to OMG, 2005-01-01, January 2005.

[66] D.D. Gajski. *Principles of Digital Design.* Prentice Hall Inc., Eaglewood Cliffs, NJ, USA, 1997.

[67] D.D. Gajski, J. Zhu, R. Doemer, A. Gerstlauer, and S. Zhao. *SpecC: Specification Language and Methodology.* Kluwer Academic Publishers, Boston/Dordrecht/London, 2000.

[68] A. Girault, B. Lee, and E.A. Lee. Hierarchical Finite State Machines with Multiple Concurrency Models. In *IEEE Transactions on Computer-Aided Design of Integrated Circuits and Systems*, 18(6), June 1999.

[69] Ch.A. Giumale and H.J. Kahn. Information Models of VHDL. In *32nd International Design Automation Conference (DAC'95),* San Francisco, CA, USA, 1995.

[70] M. Glinz. Problems and Deficiencies of UML as a Requirement Specification Language. In *International Workshop on Software Specification*, November 2000.

[71] R. Goering. *Next-Generation Verilog Rises to Higher Abstraction Levels.* EE Times, March 15, 2002.

[72] H. Gomaa. *Designing Concurrent Distributed, and Real-Time Applications with UML.* Addison-Wesley, Boston, 2000.

[73] P.N. Green and M.D. Edwards. Platform Modelling with UML and SystemC. In *Forum on Specification and Description Languages (FDL'02)*, Marseilles, France, 2002.

[74] P.N. Green and M.D. Edwards. The Modelling of Embedded Systems Using HaSoC. In *Design Automation and Test in Europe (DATE'02)*, Paris, France, March 2002.

[75] P.N. Green and M.D. Edwards. The Modeling of Embedded Systems Using HASoC. In *Design, Automation and Test in Europe (DATE'02)*, Paris, France, March 2002.

[76] P.N. Green and S. Essa. Integrating the Synchronous Dataflow Model with UML. In *Design, Automation and Test in Europe (DATE'04),* Paris, France, 2004.

[77] P.N. Green, M.D. Edwards, and S. Essa. Enhancing UML to Support the Specification of Behavior for Embedded Systems-on-a-Chip. In *UML for SoC Design Workshop at DAC'04 (UML-SoC'04)*, San Diego, CA, USA, June 2004.

[78] P.N. Green, M.D. Edwards, and S. Essa. HASOC — Towards a New Method for System-on-a-Chip Development. In *Design Automation for Embedded Systems*, 6(4), July 2002.

[79] Th. Grötker. Modeling Software with SystemC 3.0. In *6th European SystemC Users Group Meeting,* Stresa, Italy, October 2002.

[80] Th. Grötker, S. Liao, G. Martin, and S. Swan. *System Design with SystemC.* Kluwer Academic Publishers, Boston/Dordrecht/London, 2002.

[81] Y. Ha, P. Schaumont, M. Engels, S. Vernalde, F. Potargent, L. Rijnders, and H. de Man. A Hardware Virtual Machine for the Networked Reconfiguration. In *11th IEEE International Workshop on Rapid System Prototyping (RSP 2000)*, Montreal, Canada, 2000.

[82] A. Habibi and S. Tahar. A Survey on System-on-a-Chip Design Languages. In *IEEE International Workshop on System-on-Chip for Real-Time Applications*, Calgary, Alberta, Canada, July 2003.

[83] N. Halbwachs, P. Caspi, P. Raymond, and D. Pilaud. The Synchronous Data Flow Programming Language LUSTRE. In *Proceedings of the IEEE*, 79(9), 1991.

[84] M. Hans and R.W. Schafer. *Lossless Compression of Digital Audio.* Technical Report, Client and Media Systems Laboratory, HP Laboratories Palo Alto, 1999.

[85] D. Harel. Statecharts: A Visual Formalism for Complex Systems. In *Science of Computer Programming*, 8(3), June 1987.

[86] D. Harel and R. Marelly. *Come, Let's Play. Scenario-Based Programming Using LSCs and the Play-Engine.* Springer-Verlag, Berlin, 2003.

[87] D. Harel and B. Rumpe. *Modeling Languages: Syntax, Semantics and All That Stuff, Part I: The Basic Stuff*, Faculty of Mathematics and Conputer Science, The Weizmann Insitute of Science, Israel, September 2000.

[88] T. Hasegawa. An Introduction to the UML for SoC Forum in Japan. In *UML for SoC Design Workshop at DAC'04 (UML-SoC'04)*, San Diego, CA, USA, June 2004.

[89] Ø. Haugen, B. Møller-Pederson, and Th. Weigert. Structural Modelling with UML 2.0: Classes, Interactions and State Machines. In L. Lavagno, G. Martin, and B. Selic (eds.): *UML for Real: Design of Embedded Real-Time Systems*, Kluwer Academic Publishers, Boston/Dordrecht/London, 2003.

[90] J. Henkel, Th. Benner, R. Ernst, W. Ye, N. Serafimov, and G. Glawe. COSYMA: A Software-Oriented Approach to Hardware/Software Codesign. In *Journal of Computer and Software Engineering*, 2(3), Ablex Publishing Corp., Norwood, NJ, USA , March 1994.

[91] T. Heverhagen, R. Tracht, and R. Hirschfeld. A Profile for Integrating Function Blocks into the Unified Modeling Language. In *UML 2003 Workshop on Specification and Validation of UML Models for Real Time and Embedded Systems (SVERTS 2003),* San Francisco, CA, USA, October 2003.

[92] R. Hilderink and S. Klostermann. Transaction Level Modeling of SoC Platforms Using SystemC. In *Design Automation, and Test in Europe (DATE'02)*, Paris, France, March 2002.

[93] R.R. Hurlbut. *A Survey of Approaches For Describing and Formalizing Use Cases.* Technical Report, XPT-TR-97-03, Expertech, Ltd., 1997.

[94] H.-P. Hoffmann. *UML 2.0-Based Systems Engineering Using a Model-Driven Development Approach.* I-Logix, White Paper, 2004.

[95] E. Hubbers and M. Oostdijk. Generating JML Specifications from UML State Diagrams. In *Forum on Specification and Design Languages (FDL'03)*, Frankfurt, Germany, September 2003.

[96] IEEE. Wireless LAN Medium Access Control (MAC) and Physical Layer (PHY) Specifications: High-speed Physical Layer in the 5 GHz Band, IEEE Std 802.11a, June 2003.

[97] IEEE. Wireless LAN Medium Access Control (MAC) and Physical Layer (PHY) Specifications: Further Higher Data Rate Extension in the 2.4 GHz Band, IEEE Std 802.11g, June 2003.

[98] INCOSE Webpage. What is Systems Engineering? `www.incose.org/practice/whatissystemseng.aspx`

[99] I-Logix Webpage. `www.ilogix.com`

[100] S. A. Ito, L. Carro, and R. Jacobi. Making Java Work for Microcontroller Applications. In *IEEE Design and Test*, 18(5), IEEE Press, Sept–Oct 2001.

[101] B. Jacobs, J. van den Berg, M. Huisman, M. van Berkum, U. Hensel, and H. Tews. Reasoning about (Java) Classes. In *Object-Oriented Programming, Systems, Languages and Applications (OOPSLA'98)*, Vancouver, Canada, 1998.

[102] I. Jacobson, M. Ericsson, and A. Jacobson. *The Object Advantage: Business Process Reengineering With Object Technology*. Addison-Wesley, Boston, 1995.

[103] K. John and M. Tiegelkamp. *IEC61131-3: Programming Industrial Automation Systems: Concepts and Programming Languages, Requirements for Programming Systems, Aids to Decision-Making*. Springer-Verlag, Berlin, 2001.

[104] C. Kobryn. UML 3.0 and the Future of Modeling. In *Software and System Modeling*, 3(1), Springer-Verlag, Berlin, March 2004.

[105] F. Kordon and J. Henkel. An Overview of Rapid System Prototyping Today. In *Journal on Design Automation for Embedded Systems (DAES)*, 8(4), Kluwer Academic Publisher, Boston/Dordrecht/London, December 2003.

[106] P. Kruchten. *The Rational Unified Process: An Introduction* (3rd Ed.). Addison-Wesley, Boston, 2003.

[107] K. Lahiri, A. Raghunathan, and S. Dey. Communication Architecture Based Power Management for Battery Efficient System Design. In *39th Design Automation Conference (DAC'02)*, New Orleans, LA, USA, 2002.

[108] K. Lahiri, A. Raghunathan, and S. Dey. Fast System-Level Power Profiling for Battery-Efficient System Design. In *10th International Symposium on Hardware/Software Codesign (CODES'02)*, Estes Park, Colorado, USA, 2002.

[109] M. Lajolo. IP-Based SoC Design in a C-Based Design Methodology. In *IP Based SoC Design 2003*, Grenoble, France, November 2003.

[110] M. Lajolo, A. S. Basu, and M. Prevostini. UML Specifications Towards a Codesign Environment. In *Forum on Specification and Design Languages (FDL'04)*, Lille, France, 2004.

[111] S. Lange and U. Kebschull. Virtual Hardware Byte Code as a Design Platform for Reconfigurable Embedded Systems. In *Design, Automation, and Test in Europe (DATE'03)*, Munich, Germany, 2003.

[112] C. Larman. Use-Case Model: Writing Requirements in Context. In *Applying UML and Patterns (2nd Ed.)*, Prentice Hall, Eaglewood Cliffs, NJ, USA, 2001.

[113] R. Y. W. Lau and H. J. Kahn. Information Modelling of EDIF. In *30th International Design Automation Conference (DAC'93)*, Dallas, TX, USA, 1993.

[114] L. Lavagno, G. Martin, and B. Selic. *UML for Real: Design of Embedded Real-Time Systems*. Kluwer Academic Publishers, Boston/Dordrecht/London, 2003.

[115] E. A. Lee. *Overview of the Ptolemy Project*. Technical Memorandum No. UCB/ERL M03/25. University of California, Berkeley, CA, 94720, USA, July 2, 2003.

[116] E. A. Lee and D. Messerschmitt. Synchronous Dataflow. In *Proceedings of the IEEE*, 75(9), September 1987.

[117] F. Legrand, I. Bucaille ans S. Héthuin, L. De Nardis, G. Giancola, M.-G. Di Benedetto, L. Blazevic, and P. Rouzet. U.C.A.N.'s Ultra Wide Band System: MAC and Routing Protocols. In *International Workshop on Ultra Wideband Systems*, June 2003.

[118] C. K. Lennard, P. Schaumont, G. de Jong, A. Haverinen, and P. Hardee. Standards for System-Level Design: Practical Reality or Solution in

Search of a Question? In *Design, Automation and Test in Europe (DATE' 00)*, Paris, France, 2000.

[119] R. Leupers and P. Marwedel. *Retargetable Compiler Technology for Embedded Systems.* Kluwer Academic Publishers, Boston/Dordrecht/London, 2001.

[120] S. Lilly. Use Case Pitfalls: Top 10 Problems from Real Projects Using Use Cases. In *Proceedings of Technology of Object Oriented Languages and Systems*, August 1999.

[121] Y.C. Lin, C.C. Yang, M.Y. Hwang, and Y.T. Chang. Simulation and Experimental Verification of Micro Polymerase Chain Reaction Chip. In *Fifth International Conference on Modeling and Simulation of Microsystems (MSM 2000)*, San Juan, Puerto Rico, April 2000.

[122] T. Lindholm and F. Yellin. *Java Virtual Machine Specification.* Second Edition, Addison-Wesley, Boston, 1999.

[123] K. Marent. SoC++: A Unified Design Method from Concept to Implementation. In *TechOnLine* Journal, September 18, 2000. `www.techonline.com`

[124] G. Martin. SystemC and The Future of Design Languages: Opportunities for Users and Research In *The 16th Symposium on Integrated Circuits and Systems Design.* Sao Paulo, Brazil, 2003.

[125] G. Martin. UML for Embedded Systems Specification and Design: Motivation and Overview. In *Design, Automation and Test in Europe (DATE' 02)*, Paris, France, March 2002.

[126] G. Martin, L. Lavagno, and J. Louis-Guerin. Embedded UML: A Merger of Real-time UML and Co-design. In *Ninth International Symposium on Hardware/Software Co-Design (CODES'01)*, Copenhagen, Denmark, 2001.

[127] G. Martin and Ch. Lennard. Improving Embedded SW Design and Integration for SOCs. In *Custom Integrated Circuits Conference*, Orlando, FL, USA, May 2000.

[128] P. Marwedel. *Embedded Systems Design.* Kluwer Academic Publishers, Boston/Dordrecht/London, 2003.

[129] A. Massa. *Embedded Software Development with eCos.* Prentice Hall, Eaglewood Cliffs, NJ, USA, 2002.

[130] W.E. McUmber and B.H.C. Cheng. UML-Based Analysis of Embedded Systems Using a Mapping to VHDL. In *IEEE International Symposium on High-Assurance Systems Engineering HASE'99*, IEEE CS Press, November 1999.

[131] S. Mellor and M. Balcer. *Executable UML: A Foundation for Model Driven Architecture*. Addison-Wesley, Boston, 2002.

[132] S. Mellor, K. Scott, A. Uhl, and D. Weise. *MDA Distilled: Principles of Model-Driven Development*. Addison-Wesley, Boston, 2004.

[133] Mentor Graphics. Mentor's Application Specific Assistant Processor. `www.mentor.com/asap`

[134] Microelectronic Embedded Systems Laboratory (MESL), University of California, San Diego. *SPARK: High-Level Synthesis using Parallelizing Compiler Techniques*. April 2004. `mesl.ucsd.edu/spark`

[135] MIPS Technologies, Inc. MIPS Webpage. `www.mips.com`

[136] D. Morris, D.G. Evans, P.N. Green, and C.J. Theaker. *Object-Oriented Computer Systems Engineering*. Springer-Verlag, Berlin, 1996.

[137] E. Moser and W. Nebel. Case Study: System Model of Crane and Embedded Control. In *Design, Automation and Test in Europe (DATE'99)*, Munich, Germany, March 1999.

[138] S.S. Muchnick. *Advanced Compiler Design and Implementation*. Morgan Kaufmann Publishers, 1997.

[139] W. Mueller and G. Martin. *UML for SoC Design Workshop at DAC'04 (UML-SoC'04)*, San Diego, CA, June 2004. `www.c-lab.de/uml-soc`

[140] A. Nacul and T. Givargis. Code Partitioning for Synthesis of Embedded Applications with Phantom. In *International Conference on Computer Aided Design (ICCAD'04)*, San Jose, CA, USA, November 2004.

[141] T. Nakata, A. Matsuda, M. Shoji, S. Kuwamura, and Q. Zhu. An Object-Oriented Design Process for System-on-Chip Using UML. In *15th International Symposium on System Synthesis,* Kyoto, Japan, October 2002.

[142] S. Narayan, F. Vahid, and D. Gajski. System Specification with the SpecCharts Language. In *IEEE Design & Test*, 9(4), October 1992.

[143] K.D. Nguyen, Z. Sun, P.S. Thiagarajan, W.F. Wong. Model-Driven SoC Design Via Executable UML to SystemC, In *25th International Real-Time Systems Symposium (RTSS'04)*, Lisbon, Portugal, December 2004.

[144] Object Management Group (OMG) Webpage. `www.omg.org`

[145] Object Management Group (OMG). OMG XML Metadata Interchange (XMI) Specification, Version 1.2, OMG Specification, 2000. `www.omg.org/xml`

[146] Object Management Group (OMG). *OMG Unified Modelling Language Specification (Action Semantics)*. Version 1.4, January 2002.

[147] Object Management Group (OMG). *Model Driven Architecture (MDA)*. ormsc/2001-07-01, July 2001.

[148] Object Management Group (OMG). *Action Semantics for the UML*. ad/2001-08-04, August 2001.

[149] Object Management Group (OMG). *Meta-Object Facility (MOF)*. formal/2001-11-02, November 2001.

[150] Object Management Group (OMG). *Unified Modelling Language Specification Version 1.5*. OMG Adopted Specification, formal/03-03-01, March 2003.

[151] Object Management Group (OMG). *Platform Independent Model (PIM) and Platform Specific Model (PSM) for Super Distributed Objects (SDO) Specification*. dtc/03-09-01, OMG Adopted Specification, September 2003.

[152] Object Management Group (OMG). *MOF Model to Text Transformation RfP*, April 2004. `www.omg.org/cgi-bin/doc?ad/04-04-07`

[153] Object Management Group (OMG). *Human-Usable Textual Notation (HUTN) Specification*. formal/04-08-01, OMG Adopted Specification, August 2004.

[154] Object Management Group (OMG). *UML 2.0 Superstructure Specification*. Revised Final Adopted Specification, ptc/04-10-02, October 2004.

[155] Object Management Group (OMG). *Unified Modeling Language: Diagram Interchange Version 2.0*. OMG Draft Adopted Specification, ptc/03-07-03, July 2003.

[156] Object Management Group (OMG). *UML 2.0 OCL Specification*. OMG Adopted Specification, ptc/03-10-14, October 2003.

[157] Object Management Group (OMG). *UML 2.0 Infrastructure Specification*. OMG Adopted Specification, ptc/03-09-15, August 2003.

[158] Object Management Group (OMG). *UML Profile for Modeling Quality of Service and Fault Tolerance Characteristics & Mechanisms.* Final Adopted Specification, ptc/04-09-01, September 2004.

[159] Object Management Group (OMG). *UML Profile for Schedulability, Performance, and Time Specification.* OMG Adopted Specification, formal/03-09-01, September 2003.

[160] Object Management Group (OMG). *UML Testing Profile, Proposed Specification.* OMG document ptc/04-04-02, April 2004.

[161] Open SystemC Initiative (OSCI). *SystemC Version 2.0 beta-1 User's Guide*, 2001. www.systemc.org

[162] Open SystemC Initiative (OSCI). Functional Specification for SystemC 2.0, 2001. www.systemc.org

[163] S. Pasricha. Transaction Level Modeling of SoC with SystemC 2.0. In *Synopsys User Group Conference (SNUG'02),* Bangalore, India, May 2002.

[164] M. Pauwels, Y. Vanderperren, G. Sonck, P. van Oostende, W. Dehaene, and T. Moore. A Design Methodology for the Development of a Complex System-on-Chip Using UML and Executable System Models. In E. Villar and J. Mermet (eds.): *System Specification and Design Languages: Best of FDL 2002*, Kluwer Academic Publishers, Boston/Dordrecht/London, 2003.

[165] S. Pees, A. Hoffmann, V. Zivojnovic, and H. Meyr. Lisa Machine Description Language for Cycle-Accurate Models of Programmable DSP Architectures. In *36th Design Automation Conference (DAC'99)*, New Orleans, LA, USA, 1999.

[166] Project Technology Webpage. www.projtech.com

[167] W. Qin and S. Malik. Architecture Description Languages for Retargetable Compilation. In Y. N. Srikant and P. Shankar (eds.): *The Compiler Design Handbook—Optimizations and Machine Code Generation*, CRC Press, 2003.

[168] C. Raistrick. Developing Embedded Systems Using Model-Driven Architecture and Executable UML. In *Forum on Specification and Design Languages (FDL'02)*, Marseilles, France, 2002.

[169] C. Raistrick, P. Francis, and J. Wright. *Model Driven Architecture with Executable UML.* Cambridge University Press, 2004.

[170] C. Reichmann, D. Gebauer, and K. Müller-Glaser. Model Level Coupling of Heterogeneous Embedded Systems. In *10th IEEE Real-Time and Embedded Technology and Applications Symposium*, Toronto, Canada, May 2004.

[171] W. Reisig. *Petri Nets: An Introduction*. Springer-Verlag, Berlin, 1985.

[172] Rhapsody Webpage. `www.ilogix.com/rhapsody/rhapsody.cfm`

[173] E. Riccobene, A. Rosti, and P. Scandurra. Improving SoC Design Flow by Means of MDA and UML Profiles. In *3rd UML Workshop in Software Model Engineering (WiSME'2004)*, Lisbon, Portugal, October 2004.

[174] H. Riedel. *Design and Implementation of a Run-Time Environment for RTR-Systems*. Diploma Thesis, University of Applied Sciences Mittweida, Germany, January 2004.

[175] L. Rioux, T. Saunier, S. Gerard, A. Radermacher, R. de Simone, T. Gautier, Y. Sorel, J. Forget, J.-L. Dekeyser, A. Cuccuru, C. Dumoulin, and C. Andre. MARTE: A New Profile RFP for the Modeling and Analysis of Real-Time Embedded Systems. In *UML for SoC Design Workshop at DAC'05 (UML-SoC'05)*, Anaheim, June 2005. `www.c-lab.de/uml-soc`

[176] P. Rong and M. Pedram. Extending the Lifetime of a Network of Battery-Powered Mobile Devices by Remote Processing: a Markovian Decision-based Approach. In *40th Design Automation Conference (DAC'03)*, Anaheim, CA, USA, 2003.

[177] J. Rowson and A. Sangiovanni-Vincentelli. Interface-Based Design. In *34th Design Automation Conference (DAC'97)*, Anaheim, CA, USA, 1997.

[178] J. Rumbaugh, I. Jacobson, and G. Booch. *The Unified Modeling Language Reference Manual*. Addison-Wesley, Boston, 1998.

[179] M. Saksena, A. Ptak, P. Freedman, and P. Rodziewicz. *Schedulability Analysis for Automated Implementations of Real-Time Object-Oriented Models*. In *IEEE Real-Time Systems Symposium (RTSS'98)*, Madrid, Spain, 1998.

[180] T. Sakurai. Perspectives on Power-Aware Electronics. In *Digest of IEEE International Solid-State Circuits Conference*, February 2003.

[181] A. Sangiovanni-Vincentelli. *Defining Platform-based Design*. EEDesign of EETimes, February 2002.

[182] A. Sardini. SoC Design with UML and SystemC. In *6th European SystemC Users Meeting*, Stresa, Italy, October 2002. www-ti.informatik.uni-tuebingen.de/~systemc

[183] SAX Website. www.saxproject.org

[184] T. Schattkowsky and A. Rettberg. UML for FPGA Synthesis. In *UML for SoC Design Workshop at DAC'04 (UML-SoC'04)*, San Diego, CA, USA, June 2004.

[185] T. Schattkowsky, W. Mueller, and A. Rettberg. A Model-Based Approach for Executable Specifications on Reconfigurable Hardware. In *Design, Automation and Test in Europe (DATE'05)*, Munich, Germany, 2005.

[186] T. Schattkowsky and W. Mueller. Model-Based Design of Embedded Systems. In *7th IEEE International Symposium on Object-oriented Realtime distributed Computing (ISORC'04)*, Vienna, Austria, 2004.

[187] T. Schattkowsky and W. Mueller. Model-Based Specification and Execution of Embedded Real time Systems. In *Design, Automation and Test in Europe (DATE'04)*, Paris, France, 2004.

[188] C. Schulz-Key, M. Winterholer, T. Schweizer T. Kuhn, and W. Rosenstiel. Object-Oriented Modeling and Synthesis of SystemC Specifications. In *Asia South Pacific Design Automation Conference (ASP-DAC'04)*, Yokohama, Japan, 2004.

[189] K. Scott. *Fast Track UML 2.0.* Apress, 2004.

[190] B. Selic. Using UML for Modeling Complex Real-Time Systems. In *ACM SIGPLAN Workshop on Languages, Compilers, and Tools for Embedded Systems (LCTES'98)*, Montreal, Canada, June 1998.

[191] B. Selic. Models, Software Models and UML. In L. Lavagno, G. Martin, and B. Selic (eds.): *UML for Real: Design of Embedded Real-Time Systems*, Kluwer Academic Publishers, 2003.

[192] B. Selic. Turning Clockwise: Using UML in the Real-Time Domain. In *Communications of the ACM*, 42(10), October 1999.

[193] B. Selic, G. Gullekson, and P. T. Ward. *Real-Time Object-Oriented Modeling*. John Wiley, 1994.

[194] B. Selic and J. Rumbaugh. *Using UML for Modeling Complex Real-Time Systems*. ObjecTime Limited/Rational Software White Paper, 1998.

[195] M. Sgroi, L. Lavagno, and A. Sangiovanni-Vincentelli. Formal Models for Embedded System Design. In *IEEE Design & Test*, 17(2), June 2000.

[196] S. Shlaer and S. Mellor. *Object Lifecycles – Modeling the World in States.* Yourdon Press, Prentice Hall, Eaglewood Cliffs, NJ, USA, 1992.

[197] Silicore and OPENCORES.ORG. *Specification for the: WISHBONE System-on-Chip (SoC) Interconnection Architecture for Portable IP Cores.* Revision B.3, September 2002. `www.opencores.org`

[198] A. Sinha and A. Chandrakasan. JouleTrack – A Web Based Tool for Software Energy Profiling. In *38th Design Automation Conference (DAC'01)*, Las Vegas, NV, USA, 2001.

[199] V. Sinha, F. Doucet, C. Siska, and R. Gupta. YAML: A Tool for Hardware Design Visualization and Capture. In *13th International Symposium on System Synthesis (ISSS'00),* Madrid, Spain, 2000.

[200] SpecC Technology Open Consortium (STOC) Webpage. `www.specc.org`.

[201] L. Starr. *Executable UML How to Build Class Models.* Prentice Hall, Eaglewood Cliffs, NJ, USA, 2001.

[202] L. Starr. *Executable UML: The Models are the Code, A Case Study.* Model Integration Llc, 2001.

[203] B. Steinbach, D. Fröhlich, and Th. Beierlein. UML-Based Codesign for Run-Time Reconfigurable Architectures. In Ch. Grimm (ed.): *Languages for System Specification,* Kluwer Academic Publishers, Boston/Dordrecht/London, 2004.

[204] P. Stevens and R. Pooley. *Using UML: Software Engineering with Objects and Components.* Addison-Wesley, Boston, 2000.

[205] Sun Microsystems, Inc. *The Java Language Specification,* Second Edition, 2000.

[206] Synfora Webpage. `www.synfora.com`

[207] SysML Partners. *Systems Modeling Language (SysML) Specification, version 0.9 (DRAFT)*, January 10, 2004.

[208] Synopsys. *Synopsys CoCentric SystemC Compiler Behavioral User and Modeling Guide,* 2001.

[209] Synopsys. *DesignWare Building Block IP User Guide.* 2004. `www.synopsys.com/products/designware/docs/doc/dwf/manuals/dwug.pdf`

[210] SystemC Webpage. www.systemc.org

[211] W.H. Tan, P.S. Thiagarajan, W.F. Wong, Y. Zhu, and S.K. Pilakkat. Synthesizable SystemC Code from UML Models. In *UML for SoC Design Workshop at DAC'04 (UML-SoC'04)*, San Diego, CA, USA, June 2004.

[212] K. Tasie-Amadi and P. N. Green. Establishing the Correctness of Embedded Software. *Seminar Digest, IEE Postgraduate Seminar on System-on-Chip Design Test and Technology*, Loughborough, UK, 2004.

[213] Texas Instruments. *Programmable Double Biquad Filter for Tone Detection on Fixed Point DSPs*. Application Report SPRA482, February 1999.

[214] The Embedded Linux Consortium Webpage. www.embedded-linux.org

[215] The Open Group and IEEE. IEEE Std 1003.1, 2004. www.opengroup.org/onlinepubs/009695399/toc.htm

[216] V. Tiwari, S. Malik, and A. Wolfe. Power Analysis of Embedded Software: A First Step Towards Software Power Minimization. In *IEEE Transactions on Very Large Scale Integration (VLSI) Systems*, 2(4), 1994.

[217] UML Webpage. www.uml.org

[218] A. Vachoux, C. Grimm, and K. Einwich. Towards Analog and Mixed-Signal SoC Design with SystemC-AMS. In *IEEE International Workshop on Electronic Design, Test and Applications (DELTA'04)*, Perth, Australia, January 2004.

[219] Y. Vanderperren, M. Pauwels, W. Dehaene, A. Berna, and F. Özdemir. A SystemC Based System On Chip Modelling and Design Methodology. In W. Mueller, W. Rosenstiel, J. Ruf (eds.): *SystemC: Methodologies and Applications*, Kluwer Academic Publishers, Boston/Dordrecht/London, 2003.

[220] Y. Vanderperren and W. Dehaene. UML for SoC: One More Language at System Level? In *DAK Forum*, Trondheim, Norway, October 2004.

[221] Y. Vanderperren, G. Sonck, P. Oostende, M. Pauwels, W. Dehaene, and T. Moore. A Design Methodology for the Development of a Complex System-On-Chip using UML and Executable System Models. In *Forum on Specification and Design Languages (FDL'02)*, Marseilles, France, 2002.

[222] S. Vauttier, M. Magnan, and C. Oussalah. Extended Specification of Composite Objects in UML. *Journal of Object Oriented Programming*, 12(2), 1999.

[223] C. Veith, K. Buchenreider, and A. Pytell. Mapping Statechart Models onto an FPGA-based ASIP Architecture. In *European Design Automation Conference (EURODAC'96)*, Geneva, Switzerland, 1996.

[224] Velocity Home Page. `jakarta.apache.org/velocity`

[225] W3C. XSL Transformations (XSLT) Version 1.0., W3C Recommendation, November 1999.

[226] K. Wakabayashi and T. Okamoto. C-Based SoC Design Flow and EDA Tools: An ASIC and System Vendor Perspective. In *IEEE Trans. Computer-Aided Design*, 19(12), December 2000.

[227] Z. Wang. Fast Algorithms for the Discrete W Transform and for the Discrete Fourier Transform. In *IEEE Transactions on Acoustics, Speech, and Signal Processing,* vol. ASSP-32, August 1984.

[228] M.A. Wehrmeister, L.B. Becker, and C.E. Pereira. Optimizing Real-Time Embedded Systems Development Using a RTSJ-based API. In *Workshop On Java Technologies For Real-Time And Embedded Systems.* Lecture Notes in Computer Science 3292, Springer-Verlag, Heidelberg, 2004.

[229] M. Williamson. *Synthesis of Parallel Hardware Implementations from Synchronous Dataflow Graph Specifications.* Ph.D. Thesis, University of California at Berkeley, 1998.

[230] D. Wu, B. Al-Hashimi, and P. Eles. Scheduling and Mapping of Conditional Task Graphs for the Synthesis of Low Power Embedded Systems. In *Design, Automation and Test in Europe (DATE'03)*, Munich, Germany, 2003.

[231] Xilinx Inc. *Forge Compiler for High-Level Language Design.* 2003. `www.xilinx.com/ise/advanced/forge.htm`

[232] Xilinx. *Xilinx ISE.* `www.xilinx.com`

[233] P. Yang, C. Wong, P. Marchal, F. Catthoor, D. Desmet, D. Verkest, and R. Lauwereins. Energy-Aware Runtime Scheduling for Embedded-Multiprocessor SoCs. In *IEEE Design & Test*, 18(5), Sept–Oct, 2001.

[234] S. Yoo, G. Nicolescu, D. Lyonnard, A. Baghdadi, and A. A. Jerraya. A Generic Wrapper Architecture for Multi-Processor SoC Cosimulation

and Design. In *Ninth International Symposium on Hardware/Software Codesign (CODES'01)*, Copenhagen, Denmark, 2001.

[235] J. Zhu, D.D. Gajski, and R. Doemer. Syntax and Semantics of the SpecC+ Language. In *SASIMI Workshop*, Nara, Japan, 1997.

[236] Q. Zhu, A. Matsuda, S. Kuwamura, T. Nakata, and M. Shoji. An Object-Oriented Design Process for System-on-Chip Using UML. In *Digest of IEEE International Solid-State Circuits Conference*, October 2002.

[237] V. Zivojnovic, S. Pees, and H. Meyr. Lisa - Machine Description Language and Generic Machine Model for Hw/Sw Co-Design. In *IEEE Workshop on VLSI Signal Processing*, San Francisco, CA, USA, 1996.